Manfred Drosg

Dealing with Uncertainties

Manfred Drosg

Dealing with Uncertainties

A Guide to Error Analysis

With 24 Figures

Springer

Univ. Prof. Dr. Manfred Drosg
Universität Wien
Strudlhofgasse 4
A-1090 Wien, Austria
Manfred.Drosg@univie.ac.at

Library of Congress Control Number: 2006923231

ISBN-10 3-540-29606-9 Springer Berlin Heidelberg New York
ISBN-13 978-3-540-29606-5 Springer Berlin Heidelberg New York

Springer is a part of Springer Science+Business Media.

springeronline.com

Typesetting: Data prepared by the Author, by SPI Kolam
and by LE-TeX Jelonek, Schmidt & Vöckler GbR, Leipzig
Cover design: *design & production* GmbH, Heidelberg

Printed on acid-free paper 68/3100/YL – 5 4 3 2 1 0

I dedicate this book to my American friends,
in particular to those I met in New Mexico,
and to Peter Weinzierl,
who patronized me by arranging for me,
with the magnificent support
of R.F. Taschek, former P-Division leader of LANL,
to become the first foreign postdoctoral fellow at LANL.

Preface

For many people uncertainty as it occurs in the scientific context is still a matter of speculation and frustration. One of the reasons is that there are several ways of approaching this subject, depending upon the starting point. The theoretical part has been well established over centuries. However, the application of this knowledge on empirical data, freshly produced (e.g., by an experiment) or when evaluating data, can often present a problem. In some cases this is triggered by the word "error" that is an alternative term for uncertainty. For many the word error means something that is wrong. However, as will be shown, an uncertainty is just one characteristic of scientific data and does not indicate that these data are wrong. To avoid any association with something being wrong, the term error is avoided in this book whenever possible, and the term uncertainty is used instead. This appears to be in agreement with the general tendency in modern science.

The philosopher Sir Karl Popper made it clear that any scientific truth is uncertain. Usually, uncertainty is mentally associated only with measured data for which an "error analysis" is mandatory, as many know. This makes people think that uncertainty has only to do with measurements. However, all scientific truths, even predictions of theories and of computer models, should be assigned uncertainties. Whereas the uncertainty of measured data is rather easy to determine, it is too difficult, if not impossible, to establish reliable uncertainties for theoretical data of either origin.

Thus this book deals mainly with uncertainties of empirical data, even if much of it is applicable in a more general way. In particular, I want to promote a deeper understanding of the phenomenon of uncertainty and to remove at least two major hurdles en route. One is to emphasize the existence of internal uncertainties. Usually only external uncertainties are considered because they are the direct result of the theoretical approach. The former are the result of a deductive approach to uncertainties, whereas the latter are obtained inductively. The other hurdle is the so-called systematic error. This term is not used unambiguously, giving cause to many misunderstandings. It is used both for correlated (or systematic) uncertainties and for systematic deviations of data. The latter just means that these data are wrong, that is, that they should have been corrected for that deviation. There are even books in which both meanings are intermingled!

Not using the term error will make such misconceptions less likely. So I speak of uncorrelated uncertainty instead of random error, and of correlated uncertainty (and of systematic deviation, respectively) instead of systematic error. In addition, it will be shown that these two types of uncertainties are of the same nature. Thus a remark taken from a more recent book like "there is no evidence that you cannot treat random and systematic errors the same way" is self-evident.

My first interest in the subject of this book goes back to 1969, when Nelson (Bill) Jarmie at Los Alamos National Laboratory, USA, who was a pioneer in accurate measurements of cross sections, introduced me to various subtleties in this field. I am indebted to him for many insights. Not surprisingly, quite a few examples deal with nuclear physics. In this field (and in electronics) I am most experienced and, even more important, uncertainties of data based on radioactive decay can easily be determined both deductively and inductively.

The essence of this book is found already in work sheets that I prepared for undergraduate students in an advanced practical physics course when it became clear that nothing like it was available in either German or English books. This lack is the reason for not including a reference list.

Students and colleagues have contributed by asking the right questions, my colleague Prof. Gerhard Winkler by way of enlightening discussions and very valuable suggestions and M.M. Steurer, MS, by reporting a couple of mistakes. My sincere thanks to all of them.

I sincerely urge my readers to contact me at Manfred.Drosg@univie.ac.at whenever they can report a mistake or want to suggest some additional topic to be included in this book. Any such corrections or additions I will post at http://homepage.univie.ac.at/Manfred.Drosg/uncertaintybook.htm.

Vienna, September 2006 *Manfred Drosg*

Foreword by the Translator

My first contact with the topic "uncertainties" dates back to my first practical physics course at the university. The theory and practical procedure were not explained very well. I was quite confused, so I asked my dad (M. Drosg) to explain it to me, and that helped! Now, several years later, he asked me to translate the German version of this book into English to make the answers to those questions that bugged me (early on in my studies) available to a greater number of people. This was a great idea and quite a challenge for me! Although I am US-born, I spent only a little time in American schools, but, several months at Los Alamos National Laboratory, where I worked as a summer student. My mentor during this time was Robert C. Haight, who taught me science in English—I am very thankful for this great support!

Nevertheless, the translation work was not always easy, so it was a great help that I could rely on my dad for double-checking the text, and for finding the correct technical term when I was not sure. Although the aim was a full and correct translation of the German original, it is not unlikely that a few mistakes escaped the multiple proofreadings. I apologize for that.

In particular, I want to thank Alice C. Wynne, Albuquerque, New Mexico, a long-time family friend, for her thorough proofreading of the manuscript.

Vienna, September 2006 *Roswitha Drosg*

Prolog. *Seven Myths in Error Analysis*

Myth 1. *Random errors can always be determined by repeating measurements under identical conditions.*
Although we have shown in one case (Problem 6.3.) that the inductive and the deductive method provide practically the same random errors, this statement is true only for time-related random errors (Sect. 6.2.5).

Myth 2. *Systematic errors can be determined inductively.*
It should be quite obvious that it is not possible to determine the scale error from the pattern of data values (Sect. 7.2.4).

Myth 3. *Measuring is the cause of all errors.*
The standard example of random errors, measuring the count rate of radiation from a radioactive source repeatedly, is not based on measurement errors but on the intrinsic properties of radioactive sources (Sect. 6.2.1). Usually, the measurement contribution to this error is negligible.

Just as radiation hazard is most feared of all hazards because it is best understood, measurements are thought to be the intrinsic cause of errors because their errors are best understood.

Myth 4. *Counting can be done without error.*
Usually, the counted number is an integer and therefore without (rounding) error. However, the best estimate of a scientifically relevant value obtained by counting will always have an error. These errors can be very small in cases of consecutive counting, in particular of regular events, e.g., when measuring frequencies (Sect. 2.1.4).

Myth 5. *Accuracy is more important than precision.*
For single best estimates, be it a mean value or a single data value, this question does not arise because in that case there is no difference between accuracy and precision. (Think of a single shot aimed at a target, Sect. 7.6.) Generally, it is good practice to balance precision and accuracy. The actual requirements will differ from case to case.

Myth 6. *It is possible to determine the sign of an error.*
It is possible to find the signed deviation of an individual data value but the sign of the error of a best estimate, be it systematic or random, cannot be determined because the true value cannot be known (Sect. 7.2.1). The use of the term systematic error for a systematic deviation is misleading because a deviation is not an uncertainty at all.

Myth 7. *It is all right to "guess" an error.*
The uncertainty (the error) is one of the characteristics of a best estimate, just like its value, and nearly as important. Correct error analysis saves measuring time and total cost. A factual example for that is given (Sect. 10.1.1) where correct error analysis could have saved 90% of the cost.

Contents

1

Introduction

According to Richard P. Feynman, a 1965 Physics Nobel Prize winner, modern science is characterized by uncertainty. In his talk at the National Academy of Sciences in 1955 he put it this way: "Scientific knowledge is a body of statements of varying degree of certainty – some most unsure, some nearly sure, but none absolutely certain." (As found in Feynman, RP (1997) *Surely You Are Joking, Mr. Feynman*, Norton, New York)

Sir Karl Popper uses the same idea in his book "Logik der Forschung" (Popper, KR (2002) *The Logic of Scientific Discovery*, Routledge, London). There he states that scientific truth is *always* uncertain.

This thought might be frustrating, especially for beginners, but we should get used it. As Feynman says: "Now, we scientists are used to this, and we take it for granted that *it is perfectly consistent to be unsure, that it is possible to live and NOT know*."

In a nutshell, this means that there can be no scientifically relevant data *without uncertainty*. If we look closely at this last sentence we find the following two truths:

1. All scientifically relevant data have an uncertainty.
2. Data without uncertainty cannot be relevant scientifically.

Contrary to general belief, uncertainties are not the trademark of measurements. They are the trademark of science. There are measurements without uncertainties (those done by counting consecutive events, Sect. 2.1.4), but, up to now, no scientific fact without uncertainty has been found. Probably, this misconception that the measurement "makes" the uncertainty originates from the fact that only the uncertainty of (empirical) data can easily be dealt with. For the same reason we will concentrate on such uncertainties.

How do we deal with the inexactness of data? If a data value y has an (*absolute*) uncertainty Δy, we can get the degree of exactness by dividing Δy

by y, thus obtaining a *dimensionless* quantity, the *relative* (or fractional or percentage) uncertainty σ_r

$$\sigma_r = \Delta y / y \,. \tag{1.1}$$

That the absolute uncertainty Δy is not suited, e.g., for comparisons, we will see in the following.

Example. ***Absolute vs. Relative Uncertainty***

The lattice constant in a cubic lattice was measured to be (44.89 ± 0.10) nm. The distance between a point on the earth's surface to a certain point on the moon's surface is known to ± 1.0 m. Thus it is obvious that the quality of a measurement is not necessarily determined by the absolute uncertainties.

Problem

1.1.
(a) Compare the relative uncertainties of both measurements of the example, and determine their ratio.
(b) Compare the absolute uncertainties. What is the corresponding ratio?

1.1 The Exactness of Science

Let us consider the "exactness" of sciences. Those of you who (still) have not gotten used to the idea of "living and not knowing" (as Feynman put it) can skip to Chap. 2.

In science we try to explain reality by using models (theories). This is necessary because reality itself is too complex. So we need to come up with a model for that aspect of reality we want to understand – usually with the help of mathematics. Of course, these models or theories can only be simplifications of that part of reality we are looking at. A model can never be a perfect description of reality, and there can never be a part of reality perfectly mirroring a model. This statement might seem a little rash, but it can be "proved" by the following idea by Popper. He says that even if we find that there is no difference between reality and the model, this statement itself is uncertain.

Due to the theory that underlies uncertainties an infinite number of data values would be necessary to determine the *true value* of any quantity. In reality the number of available data values will be relatively small and thus this requirement can never be fully met; all one can get is the *best estimate* of the true value.

But the trouble already starts with a single measurement: A scientific quantity, e.g., a length, is only defined in the model and not in reality. (How can you define in nature an ideal straight line on which you can exactly define a distance?) For instance, if you are interested in measuring the length of a cylindrical pole, a straight line has first to be defined on or in this pole, so that the measurement can be conducted along this line. In a model this can easily be done, but in reality there is no unambiguous way of doing it. No pole has a perfectly round circumference, (small) bumps exist all over the length of the pole. The ends of the pole can never be perfectly even and will never be exactly at a right angle to the (symmetry) axis of the pole. (Even this axis can only be determined in the model, not in reality.) In addition to all this, we have to bear in mind that the pole is made of atoms that are in dynamic equilibrium with their surroundings. Atoms evaporate from the pole's surface, other atoms from the surrounding medium are deposited. Above all, they are oscillating.

We now could improve our definition of the length of the pole by improving our model, by describing the pole as a bent cylinder with dents and protrusions, etc., but we will never be able to reach an exactly *reproducible* definition of the length that is valid over time. This is exactly along the lines of Popper's statement on uncertainty in scientific discovery. Uncertainty is already introduced with the definition of the quantity, i.e., by using a model, even before it is measured. This is something not generally accepted because *usually uncertainty is attributed to measurements only!*

Contrary to general belief, we can claim that a measurement by itself does not inevitably result in an uncertainty. Only *results of scientifically relevant measurements* are inevitably uncertain, uncertainties being a *consequence of science.* This seems another rash statement, but it is based on the fact that consecutive counting (Sect. 2.1.4) can be done without uncertainty. Only if such a result is to be used in a scientific context is the uncertainty inevitable.

It is also inevitable for any model or theory to have an uncertainty (a difference between model and reality). Such uncertainties apply both to the numerical parameters of the model and to the inadequacy of the model as well. Because it is much harder to get a grip on these types of uncertainties, they are disregarded, usually.

The measurements conducted in an experiment are the connecting link between science (the model) and reality. We check the validity of a theory by comparing the measured results with the prediction by the theory. If they are not consistent, a new theory may evolve based on these experimental results.

Problems

1.2. Is the definition of a scientific quantity done in the framework of a theory (i.e., within a model) or within reality (nature)?

1.3. Name two categories of uncertainties in theories (models).

1.2 Data Without Uncertainty

According to Popper it would be sufficient to find one single scientific truth without uncertainty to falsify his statement that scientific truths are always uncertain. Let us look at some of the most obvious arguments a physicist might come up with in such falsification attempts:

Examples. ***Search for Exact Data***

- Is there a body of exactly 1-kg rest mass?
  Yes, the kilogram prototype that is kept in Sèvres near Paris; but it has this mass *per definition*. The interesting thing about it is that the mass of the kilogram prototype has changed in the course of the decades. This is one of the reasons why (seen from today's point of view) it is not a good reference mass.
- Counting can be done without uncertainty, so the result is an exact number. Is this not a falsification? No. It is a mere (historical) fact if at a certain point in time there were, for instance, four apples in a basket or a certain reading on an electronic counter. However, such information is not scientifically relevant. As soon as these numbers (obtained by counting) are brought into a scientific context, e.g., after introducing them into a model, they become uncertain. At any rate, an uncertainty in the "point of time" will be present (Sect. 2.1.4).
- The rest masses of elementary particles of the same type are the same. If this were not the case it would cause a lot of complications. Nevertheless, it could not (and cannot) be shown that two elementary particles of the same kind have exactly the same rest mass: two identical mass numbers with an infinite number of digits.
- When breaking up a boron-10 nucleus into its nucleons, ten particles will be detected, five of which will be identified as protons, the others as neutrons. This is a counting process and, therefore, can be exact. So a boron-10 nucleus consists of five protons and five neutrons (minus the binding energy). Is such a result without uncertainty? Theory requires that the number of nucleons be an integer; therefore, this theory is satisfied with the detection of ten particles with properties close to that of nucleons. However, it cannot be proven that these particles have exactly the properties of nucleons. Besides, it cannot be excluded from the experiment that there was an 11th fragment, a minute fraction of a nucleon that was undetectable because of its tiny mass. The nonexistence of such a fragment cannot be proven experimentally; all that can be done is to determine an upper limit for its mass. So it cannot be proven for sure (i.e., without uncertainty) that a boron-10 nucleus is made up of five protons and five neutrons, even if it is so obvious in view of the prevalent theory.
- The rest mass of a photon equals zero – which means this mass is known without uncertainty. No, this statement is only a hypothesis due to a theory

that is generally accepted. It cannot be proved, only falsified. From today's point of view we have no reason to doubt this theory, and thus we use a rest mass of zero for photons. As those of you interested in physics will remember, it was shown not too long ago that the rest mass of the neutrino is *not* zero, even if a rest mass of zero was part of a successful and well-established theory. Theories can lose their validity rather suddenly!

As a consequence, it can *never* be said without uncertainty that a (scientifically relevant) quantity equals zero (or any other whole number; see also Sect. 3.2.4). The number zero is special, as this number is independent of the units by which the quantity is given. The fact that there cannot be exact scientific knowledge, i.e., knowledge without uncertainty, can be seen as a result of Popper's statement that theories can only be falsified, but never be proved.

2 Basics on Data

The main subject of this book is uncertainties of empirical data. There exist all kinds of such data:

- There are statistical data like the average age of the population of a country. The mean of such data is just the mean and nothing else. In addition, this mean changes with time.
- There are also engineering data. Properties of mass-produced goods scatter around a nominal value. However, this nominal value is rather arbitrary and may be changed any time.
- Scientific data are special, in that they are the response of nature. *Only this type of data represents a truc value* that is supposedly unchangeable in time. Although such a true value can never be known exactly due to the uncertainty of scientific knowledge, its value can be *estimated* by measurements with varying degrees of certainty.

The main issue of this book is the best estimate of the true value and the accuracy with which it represents the true value.

For some scientific data the *true value* cannot be given by a constant or some straightforward mathematical function but by a probability distribution or an expectation value. Such data are called *probabilistic*. Even so, their true value does not change with time or place, making them distinctly different from most statistical data of everyday life.

A numerical description of the property of data requires defining a unit by which the property in question should be expressed, so that the property can be stated as a product of a number with this unit. Physics data, which are usually the result of measurements, have the advantage that they lack ambiguity and have well-defined true values due to well-developed physical theories. Consequently physics is a good playground when learning about uncertainties.

2.1 What Is a Measurement?

In a measurement a specific numerical value is assigned to the value of a physical variable of the sample under investigation. This is done by determining the ratio of the size of this variable to that of a standard. In order to do so, the quality of this variable and the units in which it shall be measured must be defined uniquely. This definition is based on some model that implies the correct measuring process. So we get

$$\text{physical quantity} = \text{numerical value} \cdot \text{unit}\,.$$

The following basic units have been put forward by the "Conférence Générale des Poids et Mesures" (the so-called SI base units):

- mass m, the kilogram (kg),
- time t, the second (s),
- length l, the meter (m),
- electric current I, the ampere (A),
- temperature T, the Kelvin (K),
- luminous intensity, I_L, the candela (cd),
- amount of substance, the mole (mol).

The primary standards of the first three of these physical quantities are:

- the kilogram prototype, a cylinder made of platinum–iridium that is kept in Sèvres near Paris, representing the mass unit of 1 kg,
- the frequency of radiation emitted due to the transition between the two levels of the hyperfine structure of the ground state in ^{133}Cs for frequency (and time) measurements, and
- the speed of light in vacuum (299,792,458 m/s) for speed (and distance) measurements.

It is obvious that the above primary standards will only be used for very few measurements. In major countries of the world national bureaus of standards provide services to make the standard values available. However, most measurements are conducted in relation to *secondary standards.*

2.1.1 The Best Estimate

Only when a measurement is brought into scientific context, i.e., when it is brought into connection with a scientific model or theory, do the measurement results become scientifically relevant. To become the *best estimate* of the true value defined in theory, the *measured* value has to be corrected and adjusted when necessary. There are cases in which there is no numerical difference between the measured value and the best estimate. In such cases one usually talks about measured values and means best estimates.

Let us think about an example to illustrate the difference between a measured result and the best estimate derived from it. It is possible that the net mass obtained as a difference of two mass measurements (e.g., the mass of a container minus the mass of the empty container) has a negative measurement value. After double-checking all possible sources of mistakes and finding none, it appears that this result is "correct". But – in accordance to the presently valid models – there are no negative masses, so the best estimate of the true mass simply cannot be negative. There is an escape: The best estimate is not given by a value but as an (upper) limit for this quantity (Sect. 3.2.6).

2.1.2 Direct Measurements (by Comparison)

In a direct measurement the measurement is done *relative to a standard of the same (physical) dimension*: two measurement values (i.e., the value for the standard and the value for the unknown sample) are compared. Such measurements, where a direct comparison is possible, can be the most exact, although it is obvious that the measurement value can never be more exact than the standard (see Sect. 8.1.4, and the ratio method, Sect. 10.5.1).

2.1.3 Indirect Measurements

Taking advantage of the highly evolved sensor and detector technology many measurements, even in everyday life, are now conducted indirectly (e.g., speedometer in a car, electronic scale, radar speeding control, and many more). In science indirect measurements have been common for quite a long time. In most cases the physical quantity of interest is converted into an electrical or optical signal that can easily be measured.

In indirect measurements a conversion factor (for linear conversions), or conversion curves (or tables) are used to transform the measured value into a value of the desired dimension.

Problems

2.1. What is really measured by the speedometer of a car?

2.2. What is the dimension of this measured value?

2.3. What is the dimension of the conversion factor necessary to get the speed?

2.1.4 Counting

Measurements done by counting differ basically from those depending on calibrations (i.e., those using reference data) because counting is based on natural units, whereas the others rely on standards that have been put forward (more or less) arbitrarily. Therefore, counting results are usually given by integers, i.e., without uncertainty. However, the situation is much more complex than it seems, so it deserves some thought. One has to consider both the counting process and the counting result. The latter is an integer number (without uncertainty). However, the counting process, which is always a sequential process, is open to several sources of uncertainties.

The Counting Process

Generally, the counting process comprises four actions:

1. checking whether there are any (more) objects to be counted,
2. identifying one of the objects to be counted,
3. labeling it,
4. incrementing the counter.

These actions take time. Thus two distinctly different cases must be considered. One is the counting of "events" on the time axis (i.e., of events that occur after each other); the other is counting a population present in some specified volume at some distinct moment.

When counting consecutive events the length of the counting interval is chosen beforehand. Thus it is just like measuring a count-rate. The counting process is straightforward and, usually, introduces hardly any uncertainty. Just the length of the time interval and its position on the time axis will be uncertain. If a counting loss (Sect. 4.1.2) correction is needed a further uncertainty occurs.

Counting some population that is present at a distinct moment is even more intricate. First of all, some procedure must be established to ensure that all elements are counted. Because counting takes time it is necessary to dilate that distinct moment to a time interval that is long enough to count all elements, i.e., to dilate it to some unpredictable length (because the number to be counted is not known). This requires that during the entire counting process the population remains unchanged. Such a knowledge can only be established by some theory that necessarily will be uncertain to some degree. Additional uncertainties will occur in the position of the "moment" on the time axis and in the boundaries of the space that is considered. Taking a (photographic) picture of the objects to be counted and counting the images of the objects defines the moment of counting better, but the exposure time, even in high-speed cameras, is not zero. Therefore basically the same problems arise. Even if the number of recorded counts has no uncertainty, every scientifically relevant statement relying on a counted result inevitably has an uncertainty.

The process of counting "zero" is aborted after the first step of the general counting process described above. So far, no scientific truths has been found that relies *exclusively* on recorded counts.

2.1.5 Indirect Counting

The number of identical components, e.g., bolts of a given size, can easily be determined by weighing both one bolt and the total number of bolts and determining the ratio of both weights. As long as the precision of these measurements is sufficient (and all bolts are of the same kind) the ratio rounded to an integer number (Sect. 2.3.3) will give the correct number. However, this is just a case of weighing and not of counting.

Of course, there are many more ways of counting indirectly; in many cases the precision of such measurements will not suffice to arrive to at an integer number without uncertainty as, e.g., in measuring the number of (singly) charged particles via the total electric charge.

2.2 Analog vs. Digital

Analog quantities appear to be continuous, while *digital* quantities are discrete. An example would be the number of children in a family (which definitely is a discrete variable) or their height (which changes continuously and is, therefore, a continuous variable).

> As can be easily seen, the result of *every* measurement is, in fact, digital because it is given by a finite number (Sect. 2.1).

The significant difference between "digital" and "analog" instruments is the way in which the readings are obtained: In the case of an analog instrument the read-out, that is the conversion of the (analog) value to a (digital) number, is done by the experimenter. With the digital instrument the conversion is done by electronics. The advantages of digital instruments lie in better resolution and the possibility of instantly processing and automatically saving the data. Additionally, reading a digital display does not require much expertise because the experimenter only has to transfer the numbers.

Yet digital instruments have disadvantagcs, too. The measurement procedure is more obscure, and a quick, rough reading is not easily possible. (Just think about the difference between reading an old-fashioned and a digital watch!) Also, it is more difficult to get a first idea of trends in the data during the measurement.

Problem

2.4. Decide whether the following variables are analog or digital:

(a) charge of an accumulator,
(b) number of books on a shelf,
(c) sum of points reached after rolling four dice,
(d) actual lifetime of a technical item,
(e) number of data points obtained in a measurement series,
(f) number of shares bought daily at a stock market,
(g) speed of a car,
(h) money you owe,
(i) count rate, i.e., events per time.

2.2.1 “Analog” Measurements

In Sect. 2.1 it was said that a measurement will result in a numerical value. But aren’t there *analog measurements*, too?

It is rather usual that at some place in a house one will find marks on the wall depicting the height of one or of all children at some date(s). Surely, this must be an example of an analog measurement. At a given moment the child was just as high as the mark on the wall shows.

However, the correct interpretation of this measurement process shows that this kind of measurement is digital, too. The distance from the floor to the mark serves as a secondary standard. Then the child’s height becomes $1.0.. \pm 0.n..$ units of this secondary standard.

Another example is easier to understand. Measuring a (not too long) distance by consecutively putting one shoe infront of the other is rather common. So the width of a room might have been measured to be eleven and a half shoe lengths (with some uncertainty, of course). Obviously such a procedure is not really different from measuring something that happens to have “exactly” the length of one shoe. This is just the same situation as above.

2.3 Dealing With Data (Numerals)

Numbers can only be displayed with a finite number of digits. This is independent of the way in which the numbers are handled – by hand or with a computer. In computers the number of digits are determined by the hard- and software used; humans are more flexible. In both cases the number of digits to be used must be sufficient to produce the correct result. On the other hand, it would be a waste of resources to keep too many digits.

Later on, in Sects. 3.2.4 and 9.2, we will deal with the *significance* of digits. Here we shall look only at the different ways of presenting numerical data.

2.3.1 Valid Digits

A number that is the result of a count has no uncertainty, and therefore all digits are valid. Dealing with decimal numbers is a little more complicated. The number 10.00, for instance, has four valid digits, 10.0 has only three, and 0.0010 has only two valid digits. Another way of writing this last number using only the valid digits is 1.0×10^{-3}.

Valid digits may not be left out, even if the valid digit is zero, as in the case of 10.00.

A valid digit is not necessarily a significant digit. The *significance* of numbers is a result of its scientific context (Sect. 3.2.4).

2.3.2 Truncation of Numbers

When truncating numbers, all digits after a certain digit are *cut off* without changing the other digits, i.e., the last remaining digit. This new number has approximately the same value as the original number, but it can never be greater than the original number. Truncation is most often encountered in computers that can only deal with a certain number of digits (due to the limited number of bits used).

Example. ***Truncation***

3.14159265 can be truncated to 3.1415.

2.3.3 Rounding

Generally numbers are rounded *symmetrically*: The numbers 0, 1, 2, 3, and 4 are rounded down (equivalent to truncating), and numbers 5, 6, 7, 8, and 9 are rounded up, i.e., the last remaining digit is *incremented* by one.

Examples. ***Rounding***

- 3.14159265 can be rounded (down) to 3.14 (two remaining digits after the decimal point).
- 3.14159265 can be rounded (up) to 3.1416 (four digits after the decimal point).

When rounding numbers that have been rounded previously, the cumulative rounding may introduce an error, i.e., it will give a false result.

Example. ***Consecutive Rounding***

When rounding 2.249 to one digit after the decimal point, we get 2.2. After rounding to the second digit after the decimal point (2.25), and subsequently rounding to the first digit after the decimal point, the result is 2.3!

Rounding to the Next Even Digit

To avoid false results due to cumulative rounding, one can "round to the next even number." If the last remaining digit after symmetrically rounding up is uneven, the number is rounded down instead, making sure that the last digit is even.

Example. ***Improved Consecutive Rounding***

After directly rounding to one digit after the decimal point, 2.249 becomes 2.2. Using the alternative way of rounding as explained above, 2.249 is first rounded to two digits after the decimal point to 2.24, then after rounding again we get 2.2, the same result as obtained by rounding in one step.

Problems

2.5. Follow the routine given in the example for rounding 2.249 for the number 2.252.

2.6. The following numbers are to be rounded so that the number of digits is reduced by one:

(a) 33.6
(b) 4.848
(c) 4.84
(d) 4.8
(e) 0.056

2.7. Use scientific notation (multiples of powers of 10) and round the following numbers to one remaining digit:

(a) 75.95
(b) 75.45
(c) 366
(d) 43210

2.8. Sum up the following numbers: 1.35, 2.65, 3.95, 4.45, 5.65, 6.55, and 7.75:

(a) directly,
(b) after rounding to the first digit after the decimal point,
(c) after rounding to an even next digit after the decimal point

3 Basics on Uncertainties (Errors)

All scientifically relevant quantities must be assigned an uncertainty, as discussed earlier. This applies to all data values, not only measurement values. This uncertainty is a statistical measure of data quality. It shows how well the data, i.e., the *best estimate*, fits the (unknown) *true value*. However, it *does not specify the actual deviation between these two values.* No data value is of any use whatsoever in a scientific context without a statement on its uncertainty.

Consequently, the uncertainty is a fundamental characteristic of any scientific data value.

Thus uncertainties are probabilistic quantities (Sect. 5.4). In Sect. 5.2 theoretical aspects of uncertainties are covered. The presence of an uncertainty is indicated by the symbol $\pm$ that we used in Chap. 1 already. An interval that is symmetric on both sides of a data value and has a total length of twice the uncertainty is called the 1σ *confidence interval* (*one-sigma confidence interval*, with sigma the symbol for the *standard deviation*, Sect. 5.1.1). The *true value* is expected to lie in this interval with a probability of about 68.27%. Unless otherwise indicated, uncertainties, at least in physical science, are always these *probable uncertainties*. In addition, confidence intervals that are wider by a factor 2 or 3 are common. These are called the 2σ *confidence interval* with 95.45% confidence that the true value lies inside it, or the 3σ *confidence interval* with 99.73% confidence.

The value of an uncertainty can be *induced* (*external uncertainties*, Sect. 4.2.1) or *deduced* from the characteristics of the data value (*internal uncertainties*, Chap. 6). In the case of measurement values these characteristics also include properties of the measurement process. Obviously, they are independent of the model used to present the data values (i.e., of the mathematical function describing the result). This is why an internal uncertainty even exists for a single data value (which obviously fits into any model). *External uncertainties*, on the other hand, exist only for data sets for which a model is available (e.g., time invariance, meaning that a measured quantity does not change in time). Internal uncertainties of experimental data values depend on the experimental setup; in which way can be best shown by examples.

3.1 Typical Sources of Internal Uncertainties

A voltage between two points is measured with a voltmeter. To this end these two points are electrically connected to the input terminal of the voltmeter. The measurement value can be read from the instrument's display. (*Note:* The voltage displayed on a voltmeter is, of course, the voltage across the input terminals of the instrument!) The uncertainty of this and similar measurement results consists of three components. The sources of these are discussed in the following sections.

Even for such a simple measurement, with only three easily understood uncertainty contributions, a deeper understanding of the basic properties of uncertainties is needed (Sect. 6.2).

Note: We are only interested in the uncertainty of the measured value. To obtain the best estimate of the voltage between the two points (and its uncertainty), we must correct the measured value, taking into account the input resistance of the voltmeter and the output impedance between the two points (Sect. 7.3.2).

3.1.1 Scale Uncertainty

This uncertainty shows how well the instrument is calibrated, i.e., how well a reading at the end of the scale agrees with the value implied by the international voltage standard. The scale uncertainty is usually given as a percentage uncertainty, i.e., any value measured with this instrument has the same *percentage* uncertainty as its scale uncertainty. The actual percentage deviation of any measurement value from its true value is identical despite the fact that neither its sign nor its size is known. Such uncertainties are called fully *correlated,* as we will learn in Sect. 7.2.3.

The scale of an electronic instrument changes with time, e.g., due to thermal effects in the materials. Consequently, the scale uncertainty at the time of the calibration is not valid forever. To alleviate this situation, electronic components in accurate instruments must undergo artificial aging (thermal cycling) before calibration.

3.1.2 Nonlinearity

The interpolation uncertainty (also called nonlinearity) describes how well the calibration of the full-scale reading can be transferred to intermediate readings. If such an interpolation were strictly linear, there would be no such uncertainty.

The size of the interpolation uncertainty is usually given in percent of full scale, i.e., the interpolation uncertainty has the same *absolute* value *for all readings.* Consequently its effect is the larger the smaller the reading is. Therefore use the upper part of the operating range of an instrument to make this uncertainty small. Despite the fact that the absolute interpolation uncertainty of an instrument has the same numerical value for each measurement done with it, the deviation of the ideal interpolation from the actual one differs for different measurement values, i.e., the interpolation uncertainties of an instrument are equal in size but *not identical,* they are *not fully correlated,* as discussed in Sect. 7.2.3.

3.1.3 Digitizing Uncertainty

When measuring analog quantities, a digitizing uncertainty is inevitable – only a finite number of digits are available for the presentation of the value. Usually, both the zero point and the endpoint are necessarily off by up to 0.5 (reading) units. In digital instruments these units are called *least significant bits* (LSB). So the absolute size of this uncertainty depends on the effective resolution. In analog instruments this uncertainty is called *reading uncertainty.* The deviations due to the digitizing uncertainty are, in general, independent of each other. Therefore, digitizing uncertainties are not identical, even if they have the same value. Such uncertainties are uncorrelated(Sect. 7.2.3).

3.2 Definitions

3.2.1 Terminology

In this book we will refrain from using the term "error" as far as possible; we will use the term "uncertainty" instead, which seems to be today's tendency anyway. Probably, due to historical reasons the term error has often been preferred when speaking of uncertainties, but calling something an error evokes the association of it being "wrong". This is why, even in some books, wrong measurements and measurements with systematic uncertainties are often not differentiated. However, *uncertainties are an inevitable phenomenon in the wake of science that has nothing whatsoever to do with mistakes or wrong measurements.*

3.2.2 Necessary Requirements

It is the nature of an uncertainty that it is not known *and can never be known,* whether the best estimate is greater or less than the true value.

The *sign* of this deviation and also its *magnitude* are *unknown.* If the deviation and its sign were known, the data value could be corrected for that deviation so that no signed deviation remains for the best estimate.

3.2.3 Deviations

We need to distinguish between uncertainties and deviations – *the sign of the latter is known* ("The measured value is too large, too small!"). Deviations are often called systematic errors; however, the sign of deviations is known, e.g., when a correction has been omitted. In such a case one should call it an error, i.e., mistake, and *not* an uncertainty. It is curious that some people refer to one-time blunders as mistakes, but they call systematic blundering a systematic error that should be handled like an uncertainty.

Here we want to discuss the example of a length measurement using a yardstick. This yardstick will be worn down with time, resulting in a systematic deviation for all measurements done with it afterwards. The measured results will be wrong, i.e., too large, provided that they have not already been corrected for the yardstick's wear. The measured result should be corrected until it cannot be said whether the measured value is too large or too small. Only after this correction of the measured value can we start dealing with the uncertainties of the corrected measured value and of this correction.

> A *correction* becomes necessary if the quantity defined in theory or in a model cannot be expressed directly by the measured quantity, but shows a systematic deviation from it. Even if the *measured value is correct*, it would be wrong to use this value (i.e., without the necessary corrections) as a *best estimate* of the true value (Sects. 4.1.2 and 7.3.2).

Let us look at some examples where an omission of corrections results in a real error, but not in an uncertainty:

- mechanical wear,
- buoyancy when weighing,
- loading of electrical circuits by the measuring instrument,
- ignoring the temperature of calibration of an instrument,
- optical illusions when reading instrument displays.

Everyone is subject to optical illusions, but the "magnitude" of these illusions differs from person to person. It is very helpful, if not necessary, to come to terms with the phenomenon of optical illusions, especially for people who depend on information gathered with their eyes. A simple example for optical illusions is the following: a straight line terminated by brackets pointing inwards (><) seems longer than a straight line of the same length terminated by brackets pointing outwards (<>).

3.2.4 Random Uncertainties

We call an uncertainty *uncorrelated* (also *random* or *statistical*) if the uncertainty and all of its components do *not* depend on any of the other uncertainties or their components. Obviously, the uncertainty of a single data point can only be uncorrelated if the uncertainty has only one component. The usual example of random uncertainties occurs in counting of radioactive events (Chap. 4). It is easier to give an example when an uncertainty is *not* uncorrelated:

Two voltage measurements are conducted using the same instrument with the same range settings. Thus their scale (= calibration) uncertainty is identical, i.e., there is a 100% correlation between these two uncertainties. The other two components of the total uncertainty (Sects. 3.1.2 and 3.1.3) are not totally correlated, even if they have the same numerical value.

> If two uncertainties have the same value, this is not at all a *sufficient* requirement for a total correlation between these uncertainties (see also Chap. 7).

The uncertainty of data values can either be given *explicitly*, i.e., by stating a symmetric confidence interval about the data point (also called "error bars", Sect. 9.1.3), or *implicitly* by stating only *significant* digits.

Explicit Quotation

If a mass is given as $m = (30.000 \pm 0.004)\,\mathrm{g}$, e.g., as a result of weighing, and if we have no additional (or contradictory) information, we can assume that the uncertainty stated is the *probable uncertainty* corresponding to the 1σ uncertainty (as introduced at the beginning of this chapter). In this case this means that the "true" mass has to lie between 29.996 g and 30.004 g with a probability of 68%.

Significant Figures (Implicit Quotation)

When we come across scientifically relevant data (*best estimates*) without explicit uncertainties, we can get a rough idea of the uncertainty by examining the digits given, supposing that for the numbers given only significant figures are stated, and that no significant figures are left out. *Therefore we must always be careful to give all significant figures when presenting scientifically relevant results!*

Let us assume that the raw numbers of the above mass determination were $29.999735 \pm 0.003874\,\mathrm{g}$. From the size of the uncertainty it can be concluded that the last digit of the data value is definitely not significant. So we get the following numbers after consecutive rounding to the next even digit (Sect. 2.3.3):

- 29.99974 ± 0.00387,
- 29.9997 ± 0.0038,
- 30.000 ± 0.004, and
- 30.00 ± 0.00.

From these different ways of quoting the result, just one way, namely 30.000 ± 0.004, gives the data value with the correct number of significant figures. However, in this case the rounding uncertainty of the uncertainty is $\pm 12.5\%$ (see rounding uncertainties, below). Thus, unless the total uncertainty of the uncertainty value is noticeably larger than its rounding uncertainty (at least by a factor of 3, see Sect. 3.4.1), there are too few significant figures stated for the uncertainty. To rectify this we must add one (significant) figure to the uncertainty value, and one insignificant figure to the data value. So, if the uncertainty in this case is known to better than about 30%, the correct answer would instead be 29.9997 ± 0.0038. In Sect. 9.2 this subject is discussed further.

Taking into account the rules of rounding, stating a mass as $m = 30.00\,\text{g}$ without giving an uncertainty is equal to the explicit quotation of the uncertainty yielding $m = (30.000 \pm 0.005)\,\text{g}$. Thus, one has to assume that the *true* mass lies between 29.995 g and 30.005 g, with a probability of 68%. Some might be inclined to drop the trailing zeros and to state that $m = 30\,\text{g}$. However, integers are exact, so that no uncertainty can be assigned to such a data value. To correctly give the result with two digits one must state $m = 30.\,\text{g}$.

Special care has to be taken when doing calculations using numbers with implicit uncertainties: The result of the square of 3.5 should be given as 12, (or 12.2), but under no circumstances should we write 12.25! This is because $3.45 \times 3.45 = 11.9025$ and $3.55 \times 3.55 = 12.6025$. Here only the first digit after the decimal point should be given, if it is desirable. Thus we must keep an eye on the number of significant digits when doing any mathematical operation. For instance, the square root $\sqrt{61.34} = 7.832$, as the following two values (7.8323... and 7.8317...) can be rounded to 7.832.

The final result of a product can never have more significant digits than the factor with the smallest number of significant digits.

A self-confident scientist always pays attention to the significance of digits: When stating best estimates, whether in the experiment logbook or in publications, we should always quote numbers by giving all significant digits. Whole numbers should not be used *in connection with best estimates.* Instead of 8 we should write 8. or 8.0 or 8.00, whichever corresponds to the uncertainty of the value.

Observe that – for numbers in scientific context – zeros should never be left out, even if they are the last digit after the decimal point.

When in doubt about the "exact" number of significant figures, one should adhere to the following rule: *one figure too many is much better than one figure too few.* Dropping an insignificant figure is easy; recovering a missing significant figure without additional information is impossible.

On the Use of Integers

As pointed out, best estimates should always be given with the correct number of significant digits. How about integers in science? There are, e.g., counting results that obviously are exact integer values, but only as long as they are isolated results. As soon as they are included into a scientifically relevant framework, i.e., as soon as this result is used to establish a best estimate an uncertainty will result. Already measurement parameters like the clock time of the measurement introduce uncertainty.

Integer data values taken from established theories are a second kind of integers in science. So, the rest mass of a photon is zero per definition, but not because it is a best estimate of a measurement. Likewise, reflection indices of, e.g., (1,1,1) obtained in a diffraction experiment of X-rays in a single crystal, are not "best values", but theoretical values that fit the experimental evidence best. The same will be true for experimentally determined spin values, nucleon numbers, etc. In all these cases the use of integers is correct. These integers taken from theory are the result of an identification process based on the experimental evidence.

Uncertainties When Rounding

Let us assume that we have a result of $x = 4.$ that was rounded. What is the uncertainty that was introduced by rounding, i.e., what do we know about the "parent" number from which this value was derived? All we know is that the original number was not larger than 4.500... and not smaller than 3.500.... It could be any value in this range, and obviously we do not know the sign of the deviation from 4.! Therefore we can claim that this rounding resulted in an uncertainty of ± 0.5. (This uncertainty is random but not equally distributed, i.e., it is not normally distributed; the possible values obviously lie within a rectangular distribution.)

Problems

3.1. If integers are converted to decimal numbers they have an infinite number of significant digits. Why should best estimates not be presented by integers?

3.2. Show the result of the product $5.64 \times 3.9 = 21.996$ with significant digits only.

3.3. Add the following numbers. Give the result using only significant digits: 5.19355, 14.28, 6.9561, 11.3, and 8.472.

3.4. Multiply $27. \times 4213. \times 184.$

3.5. Convert the implicit uncertainty into an explicit statement of the uncertainty:

Implicit	Explicit quotation of uncertainty:
(a) $10^{1.}$	$\pm$
(b) $1. \times 10^1$	$\pm$
(c) 1.0×10^1	$\pm$
(d) 10.	$\pm$
(e) 10.0	$\pm$
(f) 10.000	$\pm$
(g) $1/4 (= 2^{-2.} \neq 0.25)^{\dagger)}$	$\pm$
(h) $(12.)_8^{\dagger)}$	$\pm$

$^{\dagger)}$ For ambitious readers: the implicit uncertainty is given with the base 2 or 8, but the explicit uncertainty should be given with the base 10.

3.2.5 Maximum Uncertainties (Tolerances)

In engineering, for instance, probable uncertainties lead nowhere. If the nominal diameter of an axis and its bearing were subject to a symmetric 68% confidence interval, this would result in rejecting 50% of the parts produced. Of these rejected parts one half would have too much play, and the other half could not be assembled. So it is usual to use a confidence width of 99.7% (instead of the 68% confidence width usually used in science), i.e., the 3σ instead of the 1σ confidence interval is chosen. Assuming a normal distribution (Sect. 5.2.3) means that the same absolute confidence width results in a distribution three times *narrower*. Such a confidence width is often called the maximum "error". To further improve the quality of products (especially of instruments for which a specific accuracy is guaranteed), these 0.3% of the products lying outside of the confidence width are found and discarded. This corresponds to a truncation of the uncertainty distribution on both ends. In addition, in engineering cases like above it makes sense not to use symmetric uncertainties, but instead use one-sided tolerances.

To account for worst-case situations (all deviations having the same sign) uncertainty contributions to maximum uncertainties must be added linearly rather than in quadrature (Sect. 3.4), as is done with probable uncertainties.

> As an arithmetic sum can never be smaller than a geometric one, regular addition guarantees that the correct total uncertainty is under no circumstances larger than the uncertainty calculated this way.

Therefore, manufacturers of measuring devices who have to guarantee the performance of their instrument usually suggest to add the given uncertainty components linearly.

A detailed judgment of tolerances is generally not possibly without in-depth information, which can sometimes be hard to come by, as can be seen in the following example dealing with electrical resistors.

Example. ***Electrical Resistors***

Usually, the nominal values of resistors are graded logarithmically. This means that their nominal values are produced in fixed percentage steps of, e.g., 20% or 10%, resulting in tolerances of ±10% or ±5%. This tolerance is an intrinsic property of each resistor, suggesting that the (internal) uncertainty of the resistance value has the same percentage.

There exist at least two different production processes: in the first case we get resistors of (more or less) continuous values. These resistors are then sorted into groups to fit the nominal resistor values and their tolerances. The actual values of these resistors might be distributed evenly within the rectangular distribution around the nominal value. Modern production routines are more precise and deliver resistors with very small scatter in their actual value with a mean value somewhere in the tolerance range. It is quite obvious that in such a case it is difficult (or even impossible) to calculate the probable uncertainty of the impedance in a circuit of several resistors reliably by using nominal resistance values and their tolerance. Therefore, the tolerance must be assumed to be a maximum uncertainty. This problem will be revisited in Sect. 6.2.2.

3.2.6 Limits

If zero lies inside one of the uncertainty bars of a best estimate, but zero (and values beyond zero) are not possible due to theory, it does not make sense to choose a symmetrical uncertainty interval around the best estimate. In this case the upper (or the lower) end of the uncertainty interval will be chosen as the upper (or lower) limit for the desired quantity. The best estimate will then be given as smaller (or larger) than this limit with a confidence given by that of the confidence interval used.

Example. ***Negative Net Count Rate***

In radiation measurements the (radioactive) background has to be taken into account. That is why two measurements are actually necessary: a background measurement giving N_b counts, and a foreground measurement with N_f counts (which include counts from the background and from the unknown radiation). Generally, N_f and N_b are measured consecutively, assuming that the background stays constant in time.

The values $N_f = 9918$ and $N_b = 9979$ were measured in (live) time intervals t_a of the same length (Sect. 4.1.3). The difference of -61 ± 141 counts is caused by the radiation we are trying to measure. Naturally, adding radiation from an additional source can never reduce radiation; therefore the true value of the total radiation can never be smaller than that of the background radiation. Thus, from this measurement we can only give an upper limit for the best estimate of the radiation, i.e., a 1σ limit of 80 per t_a. This means that the true value of the counts recorded in the detector (caused by the radiation we are trying to measure) is ≤ 80 per t_a with a probability of 68%.

Note: In Sect. 5.2.2 we will see that the internal uncertainty ΔN of a number N of radioactive events is $\Delta N \cong \sqrt{N}$.

3.2.7 Outliers (Flyers)

Outliers or flyers are those data points in a set that do not quite fit within the rest of the data, that agree with the model in use. The uncertainty of such an outlier is seemingly too small. The discrepancy between outliers and the model should be subject to thorough examination and should be given much thought. *Isolated data points*, i.e., data points that are at some distance from the bulk of the data are *not* outliers if their values are in agreement with the model in use.

Rarely should we encounter data points with very large deviations. In the case of normally distributed data values only one in 22 points is expected to lie outside 2σ, one in 370 points outside 3σ, and one in 15.8×10^3 points outside 4σ (Table 5.3).

Let us assume that in a set of ten data points there is one data point with a deviation of 4 standard deviations. This one data point out of ten affects the best estimate by a factor of 1.58×10^3 stronger than the expected one point out of 15.8×10^3 points would. This factor equals the ratio of the numbers of data points used in the comparison; this can be deduced from the definition of the mean value given in Sect. 4.2.1. Thus the best estimate would be overly influenced by the value of the outlier, so it is worthwhile to check this outlying value thoroughly. If there is an outlier there are two possibilities:

- The model is wrong– after all, a theory is the basis on which we decide whether a data point is an outlier (an unexpected value) or not.
- The value of the data point is wrong because of a failure of the apparatus or a human mistake.

There is a third possibility, though: The data point might not be an actual outlier, but part of a (legitimate) statistical fluctuation. Remember: In a normal distribution it is necessary for 32% of the points to deviate by more than one standard deviation. To find out if the outlier is not just a statistical fluctuation, we can use Chauvenet's criterion (Sect. 6.4.1).

Assuming that the model is faulty can lead to discoveries that can even result in the Nobel Prize. The discovery of the atomic nucleus, for instance, is the result of the correct interpretation of "outliers". Nevertheless, in most cases it will just be a waste of time if this line of thought is pursued much further.

If it appears that there is something wrong with a suspicious data point (which can be true, but does not need to be true), scientists have to accept this challenge of dealing with it, e.g., by proceeding according to the checklist below. However, the absence of "objectiveness" of such a procedure makes it quite controversial because it has the danger of tailoring the data at will.

Options of Dealing With Outliers

1. Identify and eliminate the cause of the discrepancy. Take into account the history of the data point, i.e., look for faults in the experimental setup, in the data logging, in the data reduction, and in the documentation (e.g., having written down 23.5 instead of 32.5). When checking the data and the possible sources of errors we need to examine the consistency (in particular with the help of redundant data), the completeness, and the credibility (plausibility) of the raw data. In the course of these checks it is possible to correct obvious mistakes in the data records, e.g., if it is obvious that a decimal point was lost or if (with data that were originally binary data) a factor with a (higher) power of 2 is missing. To find the cause for such mistakes the availability of detailed documentation covering all, even redundant, data will be very helpful. If the suspicious data point is an end point of some distribution, it might be that the experimenter did not understand the experimental situation well enough so that corrections have not been done correctly or have been omitted.
2. Choose a greater uncertainty for the suspicious data point.
3. Calculate the mean value of the suspicious data point with appropriate other points and adjust the uncertainty of the combined value correspondingly.
4. Exclude the suspicious data point and document the fact that you have done this.
5. Keep the suspicious data point and learn to live with the discrepancy.

Refrain from automatic checking and discarding of data, e.g., via computers, because we should always be aware of the steps taken. Discarding data is very critical. It is vital that any manipulation of data remains a rare exception and that it is well documented.

> How can discoveries occur if researchers discard data that do not fit their point of view?

Problem

3.6. One value out of five in a data set deviates by 3σ from the mean value. Regularly, such a deviation is expected for one out of 370 data points. How much stronger is the influence of this outlier on the best estimate than expected?

3.3 Uncertainty of Data Depending on One Variable

3.3.1 Length

A length L be measured by applying a yardstick of the length l k' times:

$$L = k' \cdot l \,. \tag{3.1}$$

The length l of the yardstick is known with an uncertainty of $\pm\Delta l$. What is the size of the uncertainty ΔL of the result when Δl is the only uncertainty to be considered?

Result: Any change of the length l results in a k' times stronger change of L. This dependence of L on l can also be seen from the differential coefficient:

$$\frac{\mathrm{d}L}{\mathrm{d}l} = k' \,. \tag{3.2}$$

Thus one gets

$$\Delta L = k' \cdot \Delta l = \frac{\mathrm{d}L}{\mathrm{d}l} \cdot \Delta l \,. \tag{3.3}$$

The same procedure applies for nonlinear functions as well.

3.3.2 Circular Area

The area A_C of a circle is determined from its measured diameter d. The diameter d is known within $\pm\Delta d$. By differentiating we learn how the uncertainty ΔA_C of the area depends on the uncertainty Δd of the diameter (Fig. 3.1).

The original equation

$$A_C = 0.25 \cdot \pi \cdot d^2 \tag{3.4}$$

is differentiated to give

$$\frac{\mathrm{d}Ac}{\mathrm{d}d} = 0.5 \cdot \pi \cdot d \,. \tag{3.5}$$

This equation is then transformed to give the uncertainty:

$$\Delta A_C = 0.5 \cdot \pi \cdot d \cdot \Delta d \,. \tag{3.6}$$

Inserting the uncertainty Δd into the equation for the area $\mathrm{A_C}$ (which some of you might be inclined to do) results in quite a different answer for ΔA_C:

$$\Delta A_C \neq 0.25 \cdot \pi \cdot \Delta d \cdot \Delta d \,. \tag{3.7}$$

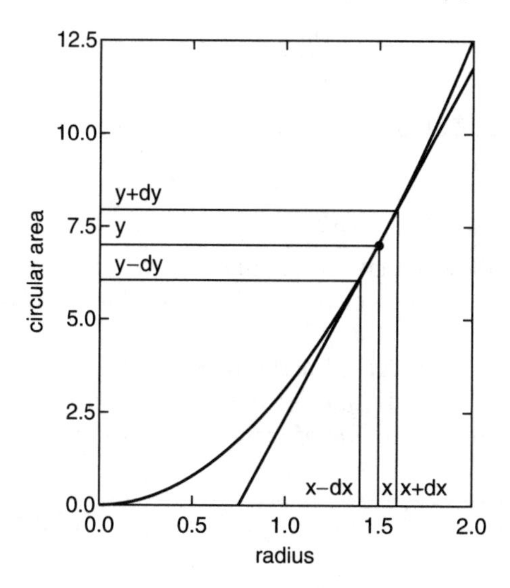

Fig. 3.1. By means of the first derivative (the tangent) we can see how a change in x affects y

3.4 Multiple Uncertainty Components (Quadratic Sum)

A velocity v is calculated using the measured values for the time $t \pm \Delta t$ and the distance $l \pm \Delta l$. The uncertainty Δv of the velocity can be expected to depend on the uncertainty Δt of the time and on the uncertainty Δl of the distance. In most cases the time measurement would be independent of the length measurement, i.e., the correlation (Chap. 7) between the uncertainties of these quantities equals zero, which means that they are uncorrelated or totally independent of each other. A correlation between these quantities could exist, for instance, if the length measurement was done indirectly by measuring a time interval. But, let us assume totally uncorrelated uncertainties.

The two uncertainty components Δv_l and Δv_t are calculated just the same way as in the previous section; from

$$v = \frac{l}{t} \tag{3.8}$$

we obtain the partial derivative:

$$\frac{\partial v}{\partial l} = \frac{l}{t}\,, \tag{3.9}$$

yielding Δv_l, the contribution of l to the total uncertainty of v:

$$\Delta v_l = \frac{\partial v}{\partial l} \cdot \Delta l = \frac{l}{t} \cdot \Delta l\,. \tag{3.10}$$

The other partial derivative equals

$$\frac{\partial v}{\partial t} = l \cdot \frac{-1}{t^2}, \tag{3.11}$$

yielding Δv_t, the contribution of t to the total uncertainty of v:

$$\Delta v_t = \frac{\partial v}{\partial t} \cdot \Delta t = l \cdot \frac{-1}{t^2} \cdot \Delta t. \tag{3.12}$$

The symbol ∂ in above equations signifies a partial derivative. When differentiating partially with respect to a certain variable, all other variables must be treated as constants.

The total uncertainty ΔF of a result $F = F(x_1, x_2, x_3, \ldots)$ is calculated by adding all n individual (*independent*) uncertainty components ΔF_{xi} in quadrature according to the *law of error propagation*:

$$\Delta F = \sqrt{(\Delta F_{x1})^2 + (\Delta F_{x2})^2 + (\Delta F_{x3})^2 + \ldots} = \sqrt{\sum_{i=1}^{n} (\partial F/\partial x_i)^2 \cdot (\Delta x_i)^2}. \tag{3.13}$$

Therefore, in the example discussed above the *absolute uncertainty* is

$$\Delta v = \sqrt{(\Delta v_l)^2 + (\Delta v_t)^2} = \sqrt{(1/t)^2 \cdot \Delta l^2 + (-l/t^2)^2 \cdot \Delta t^2}, \tag{3.14}$$

and after division by v (i.e., after multiplication with t/l) we obtain the *relative* (or *fractional* or *percentage*) *uncertainty*

$$\frac{\Delta v}{v} = \sqrt{\left(\frac{\Delta l}{l}\right)^2 + \left(\frac{\Delta t}{t}\right)^2}. \tag{3.15}$$

Note: As exemplified in Sect. 7.2.2, it is good practice to calculate the total uncertainty starting with the equation of the final result. In a step-by-step procedure using intermediate results and their uncertainties in the determination of the total uncertainty, it may happen that we miss a possible correlation between intermediate uncertainties. If a correlation between uncertainty components exists, it is not correct to use the *simple* law of error propagation. The total uncertainty would then be too small or too large, depending on the sign of the correlation. In Chap. 8 the correct way of calculating total uncertainties from their components is dealt with comprehensively.

This *quadratic sum* is also called geometric sum because the way in which this sum is calculated can be visualized by a triangle with a right angle (Fig. 3.2). The two uncertainty components are in the directions of the x-axis and the y-axis in an orthogonal coordinate system. The y-component has no contribution parallel to the x-axis, and vice versa. Thus we can see that they are uncorrelated. We get the combined uncertainty by connecting the end points of these two contributions by the hypotenuse S. The length of this hypotenuse can be calculated following the law of Pythagoras, $S^2 = A^2 + B^2$. As can be seen in Fig. 3.2, the contribution of B to the total sum S is small if

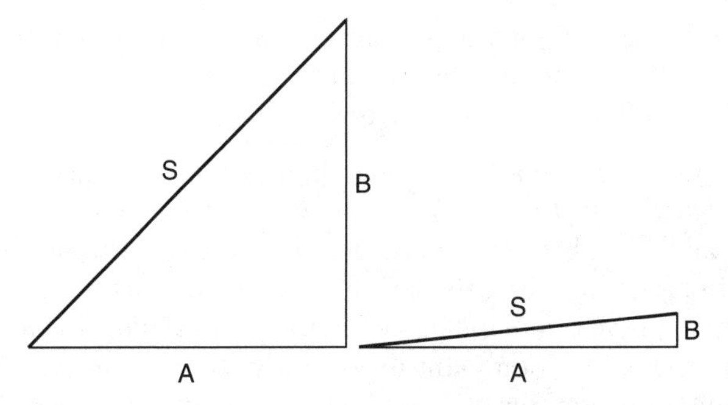

Fig. 3.2. Geometric visualization of the quadratic sum. *Left*: Both components A and B contribute equally to the sum S. *Right*: Component A is dominant. The quadratic sum S is only slightly greater than A

A is a lot larger than B. In this case we would call component A the *dominant* component.

When programming a computer for the calculation of uncertainties following the law of error propagation, its recursivity comes in very handy:

$$S_1^2 = A^2 + B^2 \,, \tag{3.16}$$

$$S^2 = A^2 + B^2 + C^2 = S_1^2 + C^2 \,. \tag{3.17}$$

Note: In the above paragraphs we have assumed (without specifically mentioning it) that the components of the uncertainties are *independent* of each other (i.e., *uncorrelated*) and that their cause is *random* (i.e., that they are *normally distributed*, see Sect. 5.2.3). In such cases it is legitimate and correct to use the quadratic sum when combining uncertainties. Using the quadratic sum gives a smaller result than adding the components arithmetically. If we have no information about the degree of correlation of the data, we should choose to use the arithmetic sum instead (e.g., in cases like in Sect. 3.2.5).

Adding uncertainty components arithmetically results in a sum of the uncertainties that can only be too large, but never too small!

3.4.1 Properties of the "Quadratic" Sum

Let us consider the following example: The length is known to ±1.00%, the time to ±3.00%, then the resulting velocity is known to $\pm\sqrt{3.00^2 + 1.00^2}\%$, which is ±3.16%. As can be easily seen, the total uncertainty of the velocity is only slightly (about 5%) greater than the uncertainty of its *dominant component*, the uncertainty of the time value. By squaring numbers, larger numbers have much more weight. If, as in our example, one number is at least 3 times greater than the other, its square is at least 9 times greater than the other number squared, so that, usually, the contribution of the smaller number can be disregarded. Then the *dominant* uncertainty contribution is a very good approximation of the combined uncertainty.

This has the following consequences:

1. Uncertainties that are a lot smaller than the largest uncertainties can be left out, *but this fact has to be documented.* ("The uncertainty of ... was ignored, as it is substantially smaller than the uncertainty of ....") This procedure facilitates the calculation of the total uncertainty, as it is only necessary to prove that small uncertainties are smaller (by at least a factor of 3) than the *largest* dominant uncertainty component, and not give their exact value (see examples in Sect. 9.4.1, estimates of uncertainties).
2. To improve the accuracy of the best estimate of a quantity that has two or more components (e.g., of the velocity in the example discussed above), it is necessary to improve the accuracy of the value with the dominant uncertainty (of the time measurement). An improvement of the other (of the length measurement) hardly has any effect at all on the total uncertainty. (Even if the length measurement is conducted with twice the accuracy, the total uncertainty will be 3.04% instead of 3.16%.) Obviously, if we want to improve the accuracy of a result, it is vital to *improve the accuracy of the component(s) with the dominant uncertainty* (Chap. 10).

Problems

3.7. Reduce the uncertainty of the time component in the example in Sect. 3.4.1 by a factor of 10. To what value is the total uncertainty (of previously ±3.16%) reduced?

3.8. The specific resistance (Ω/m) of a wire should be measured. This wire is stretched between two electrically conducting clamps. The distance between these two clamps is measured to be 0.9746 m, and the electrical resistance of the wire is 0.0053 Ω. The tape measure has a scale uncertainty of ±0.02%, an interpolation uncertainty of ±0.0003 m, and a digitizing uncertainty of ±0.0002 m. The ohmmeter (part of a $6^1/_2$-digit multimeter) has a scale uncertainty of ±0.008% and a nonlinearity of ±0.004%. The measurement is done in the 100-Ω range.

(a) What is the value of the specific resistance of the wire and its uncertainty?
(b) Which uncertainty component is dominant?
(c) Suggest ways to get a more accurate result.

3.4.2 Subtraction in Quadrature

If one of the uncertainty components, e.g., B, and the quadratic sum S are known, the other uncertainty component A can be determined as follows:

$$A^2 = S^2 - B^2 . \tag{3.18}$$

Such a procedure is needed, for example, if B' is a better value that can be used instead of B, so that a better value S' (a better the sum of the uncertainties) can be obtained from

$$S'^2 = A^2 + B'^2 = S^2 - B^2 + B'^2 . \tag{3.19}$$

This is especially important if correlated and uncorrelated (total) uncertainties have been added in quadrature to give a partially correlated total uncertainty (Sect. 8.2.3). If one component is known (or the degree of correlation is known, see Sect. 8.1), it is then possible to separate these two components via the quadratic subtraction.

4

Radioactive Decay, a Model for Random Events

To fulfill the requirements of the theory underlying uncertainties, variables with random uncertainties must be *independent* of each other and *identically distributed.* In the limiting case of an infinite number of such variables, these are called *normally distributed* (Sect. 5.2.3). However, one usually speaks of normally distributed variables even if their number is finite.

Counting of radioactive events is especially well suited for explaining problems arising in connection with uncorrelated (random) uncertainties because radioactive decay is an entirely random process and the measurement of the accompanying radiation is done by counting (Sect. 2.1.4). Remember, counting is the only type of measurement without the need of a standard. All radioactive nuclei of one kind will have the same decay probability. In addition, it is not possible for radioactive nuclei to exchange information – they cannot "know" whether another nucleus has decayed or not – so the radioactive decay of one nucleus is *independent* of all other nuclei. To ensure that the data are identically distributed, the counting time must be much smaller than the lifetime of the radioactive nuclei in question so that the change of radiation intensity during the measuring period is negligible.

4.1 Time Interval Distribution of Radioactive Events

Radioactive decay is random in time; in those cases where all nuclei have random orientation, the emission of radiation is also random in space. Both effects make radiation from radioactive sources arrive at the detector position randomly in time. As will be shown below, the frequency distribution $I(t)$ of the time intervals (between two consecutive events) can be described with the help of a probability distribution, the so-called Poisson distribution (Sect. 5.2.2).

The vertical lines on the base of Fig. 4.1 show the chronological sequence of (and the time intervals between) 25 statistical detector events recorded by a digital oscilloscope. The shortest interval between two events is 0.08 units, the

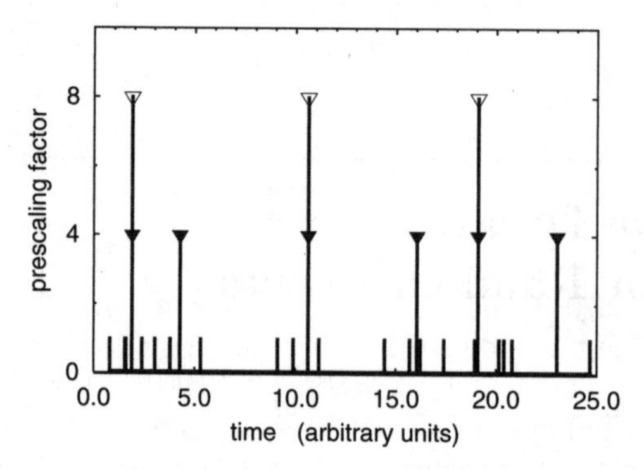

Fig. 4.1. Chronological sequence of detector signals (arbitrary section). Signals prescaled by 4 reach the height 4; those prescaled by 8 reach 8

longest amounts to 3.9 units, and the mean interval length is 1.0 units. The time structure of these events shows one of the most important characteristics of statistically arriving signals: *Short intervals* (those that are substantially shorter than the mean interval length) *occur much more often than longer intervals* (those substantially longer than the mean interval length). Short intervals are "exponentially" favored, as we can see from the frequency distribution $I(t)$ of the time intervals. This distribution gives the probability that exactly *one* event occurs in a given time interval (at the very end of it).

From the mean event rate $r = \mathrm{d}N/\mathrm{d}t$ we get the probability $\mathrm{d}N = r \cdot \mathrm{d}t$ of the occurrence of an event N in any time interval $\mathrm{d}t$, therefore the probability that in the time interval $[t, t + \mathrm{d}t]$ exactly one event will occur is $r \cdot \mathrm{d}t$. From the Poisson distribution (Sect. 5.2.2) we get the probability p_0 that within a time interval of length t no event will occur:

$$p_0 = \mathrm{e}^{-rt} . \tag{4.1}$$

The probability that no event will occur at first and the probability that an event will take place later on are independent of each other. Thus we can multiply these two probabilities and arrive at the interval length distribution:

$$I(t) \cdot \mathrm{d}t = \mathrm{e}^{-rt} \cdot r \cdot \mathrm{d}t . \tag{4.2}$$

This equation has its largest value for $t = 0$, i.e., small time intervals are most common; they are exponentially favored.

Theoretically, the interval length between two events can even be zero. In practice, very small time intervals cannot be observed because any data manipulation takes a certain amount of time, the so-called *dead time*. We will discuss this phenomenon in more detail later (Sects. 4.1.2 and 10.3).

Problem

4.1.
(a) Do all events occurring in the detector show up in Fig. 4.1 (which is the output of a digital oscilloscope)?
(b) What kind of (hypothetical) property of the digital oscilloscope would be required so that it does not lose events?

4.1.1 Prescaling

If we select every fourth (or every eighth) signal – as shown in Fig. 4.1 – this is called *prescaling* by a factor of 4 (or 8). In Fig. 4.1 the starting signal is number five.

From Table 4.1 we see that

- When dealing with random events we find intervals of much shorter length than the mean interval length (theoretically, the interval length could be zero, i.e., two events could coincide).
- The (relative) difference of the interval lengths $(\Delta t_{\mathrm{max}} - \Delta t_{\mathrm{min}})/\Delta t_{\mathrm{av}}$ becomes smaller (and smaller) after repeated prescaling. (It is not typical for this difference to be that small after prescaling by 8. This can be seen when a different starting point, e.g., the second signal and not the fifth, is chosen as starting point for the prescaling.)

The *generalized frequency distribution* $I_g(t)$ of time intervals describes prescaling by any factor of h. This distribution gives the probability that the hth signal occurs in the interval $[t, t + \mathrm{d}t]$ when the first event took place at $t = 0$ and $h - 1$ events have already occurred.

This time we get from the mean event rate $r = \mathrm{d}N/\mathrm{d}t$:

- the probability $\mathrm{d}N = r \cdot \mathrm{d}t$ that in the time interval $[t, t + \mathrm{d}t]$ exactly one event will occur, and with the help of the Poisson distribution we get
- the probability p_{h-1} that $h - 1$ events have occurred in the interval $[0, t]$:

$$p_{h-1} = \frac{(rt)^{h-1} \cdot \mathrm{e}^{-rt}}{(h-1)!} . \tag{4.3}$$

Table 4.1. Analysis of the time intervals of the 25 random signals of Fig. 4.1 (arbitrary time units)

Case	Min. space Δt_{min}	Max. space Δt_{max}	Mean space Δt_{av}	$(\Delta t_{\mathrm{max}} - \Delta t_{\mathrm{min}})/\Delta t_{\mathrm{av}}$ [%]
All data	0.08	3.90	1.00	382.
Prescaled by 4	2.30	6.45	4.23	98.
Prescaled by 8	8.45	8.75	8.60	3.5

Thus, similar to above (Sect. 4.1), the generalized frequency distribution results in

$$I_g(t) \cdot \mathrm{d}t = \frac{(rt)^{h-1} \cdot \mathrm{e}^{-rt}}{(h-1)!} \cdot r \cdot \mathrm{d}t \,. \tag{4.4}$$

Here the mean interval length t_{av} is no longer $1/r$, but is larger by a factor of h, as required for prescaling by a factor of h. Also of interest is the most probable interval length t_{mod}, the maximum of the distribution (Sect. 5.1.1)

$$t_{\mathrm{mod}} = (h-1)/r \,. \tag{4.5}$$

Not surprisingly, these (generalized) equations are true for $h = 1$, too, i.e., the case without prescaling. Then $t_{\mathrm{mod}} = 0$, i.e., small intervals are favored, as mentioned previously. For $h \gg 1$ we get $t_{\mathrm{mod}} \approx h/r = t_{\mathrm{av}}$.

The increase of the most probable interval length by prescaling is important for constructing (electronic) counters to be used for random signals. (Using just $h = 2$ results in a large change in the position of the maximum, which moves from $0/r$ to $1/r$.) In such counters it is vital that the first counter stage is especially fast to minimize the loss of counts.

Problems

4.2. Show that numerous intervals must be shortened (greatly) to maintain the total number of signals if a regular sequence of signals is disturbed by increasing the length of at least one interval markedly (e.g., by a factor of 10).

4.3. Create an alternative to Table 4.1 by choosing a different starting point (other than the fifth signal) in Fig. 4.1 for prescaling by 4 and 8. How can you explain the differences?

4.1.2 Counting Loss (Dead Time)

It cannot be avoided that some time period is spent on any kind of processing a signal. During this time no other signal can be recorded and thus gets lost. Figure 4.2 shows a portion of Fig. 4.1 that has been enlarged, and a time scale has been added.

In the example of Fig. 4.2 two signals are lost if the processing time of each (processed) signal is 5 µs. This processing time is called the *pulse pair resolution* or the *dead time* of the processor (or more accurately, dead time of the first kind, see Sect. 10.3). When counting random signals it is important that the result is corrected for lost signals by applying the so-called *dead time correction*.

In this particular example there was never more than one signal during any one dead time interval. So, a prescaling of two would suffice to eliminate the loss of counts during this particular time window. Generally, the loss of random signals due to dead time is greatly reduced if prescaling is applied (see Δt_{min} in Table 4.1). However, observe that the process of *prescaling causes dead time, too.*

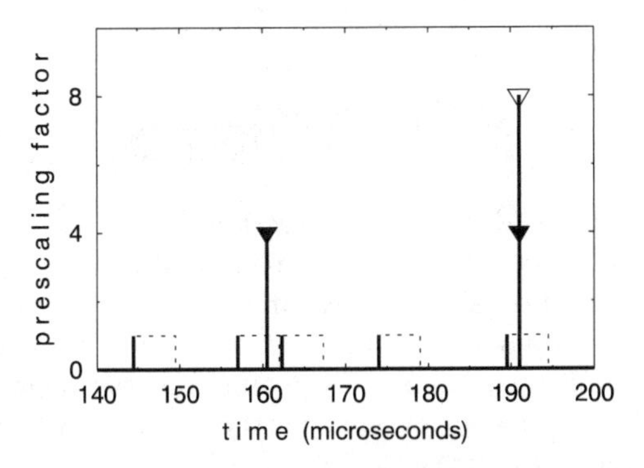

Fig. 4.2. Portion of Fig. 4.1. Loss of signals due to dead time: Events that occur during the dead time of a previous signal get lost. The dead time is indicated by *dashed rectangles* with a width of 5 μs

4.1.3 Direct Correction of Dead Time

If the time axis is divided into *intervals of equal but arbitrary length* and if the *probability of the occurrence of a signal is the same* for all intervals, we are dealing with *random signals.* That is, they arrive randomly in time.

If this is the case, the correction for signals lost due to dead time can be done directly: The time during which the system accepts signals (i.e., is not dead) is measured. During each interval of dead time the time measurement is halted, i.e., only that time is measured during which the system is "receptive" to signals. This is done with the help of a coincidence circuit interrupting the timer during each dead time interval (anticoincidence circuit). It is quite obvious that this correction is applicable independent of the length of the individual dead time interval.

The time t_a during which incoming signals can be recorded is called *live time* (active time) in contrast to the *real time*, the actual duration t_m of the measurement. From this we deduce the straightforward dead time correction factor f_{td} (for random signals)

$$f_{td} = \frac{t_m}{t_a} . \quad (4.6)$$

If N_a counts were recorded during a time interval t_m, the actual number of events N is given as

$$N = \frac{N_a \cdot t_m}{t_a} . \quad (4.7)$$

Given a mean dead time t_d per count we get

$$t_m = t_a + N_a \cdot t_d , \quad \text{and} \quad \frac{t_m}{t_a} = 1 + \frac{N_a \cdot t_d}{t_a} , \quad (4.8)$$

and the true number N of the events

$$N = N_a \cdot \left(1 + \frac{N_a \cdot t_d}{t_a}\right) = \frac{N_a}{1 - \frac{N_a \cdot t_d}{t_m}} . \tag{4.9}$$

The uncertainty of the dead time correction factor t_m/t_a stems from the uncertainties of these two time measurements. When using the same clock for these two (simultaneous) time measurements, the correlated uncertainty components, like accuracy and stability of the frequency, cancel (Sect. 8.1.4) so that the digitizing uncertainty of the time measurements dominates. By choosing the highest frequency possible for the clock the digitizing uncertainty is minimized, so that often it can be disregarded.

This live time method fails if not all signals to be counted are random and identically distributed in time. This is the case if some of the signals originate from a pulse generator of constant frequency, or if they come from radiation with such a time structure that the count rate does not appear constant in time intervals very much longer than the dead time. In such a situation the correction factor is correct only for the truly random component (with a constant count rate), but not for the other signal components.

Example. ***Dead Time but no Counting Loss***

A pulse generator set to a constant repetition frequency of 5 kHz produces a reference signal for a pulse height distribution (via the test input of the preamplifier). The dead time of the pulse processing system is constant and amounts to 10 μs. With a distance of 200 μs between the pulses of the pulse generator, the corresponding dead time of the system is 5%. Applying this conventional dead time correction to these periodic signals would give the wrong result as none of the pulses occurs during the dead time of the previous one. No periodic signal gets lost and no correction is necessary.

Even though we refer to a "*dead time* correction", its purpose is the correction *for lost counts.*

4.1.4 Correction Using a Pulse Generator

With this method we measure not the dead time, but instead the probability of counting a signal. To achieve this, signals of constant amplitude are added to the signals of the spectrum to be measured. This is done in such a way that these signals do not disturb this spectrum, e.g., by setting the pulse height so that the added pulses are the highest to be recorded. They appear at the upper end of the spectrum where no other signals are recorded. Obviously, the average period length of this generator has to be much longer than the dead time.

If N_a is the number of signals recorded in the spectrum (including the generator signals) during the measurement time t_m, and t_d is the mean dead time per signal, and N_p is the number of generator signals added during the measurement time t_m, then the line in the spectrum originating from the generator pulses will contain N_{pa} signals:

$$N_{pa} = \frac{N_p}{1 + \frac{N_a \cdot t_d}{t_a}} . \tag{4.10}$$

That is, the dead time correction factor f_{td1} giving the corrected number N is given by

$$f_{td1} = N/N_a = N_p/N_{pa} = 1 + N_a \cdot t_d/t_a . \tag{4.11}$$

Using the best estimate N_p/N_{pa} for the true ratio N/N_a is, of course, subject to a statistical uncertainty. In most cases this uncertainty will be substantially smaller than the dominant one, so it can be disregarded (Sect. 3.4). If not, it can be calculated using the binomial distribution (Sect. 5.2.1). Because the count rate stemming from the pulse generator will necessarily be small, we have $N_{pa} \ll N_a$ so that $p = N_{pa}/N_a$ and $z = N_a$. With that the absolute uncertainty ΔN_{pa} is obtained by way of the *standard deviation* σ_{bin} of the binomial distribution as

$$\Delta N_{pa} = \sigma_{\text{bin}} = \sqrt{N_{pa} \cdot (1 - N_{pa}/N_a)} \approx \sqrt{N_{pa}} . \tag{4.12}$$

Thus the fractional uncertainty of the correction factor becomes

$$\Delta f_{td1}/f_{td1} = \sqrt{N_{pa}}/N_{pa} = 1/\sqrt{N_{pa}} . \tag{4.13}$$

Periodic Signals

The validity of the above equations is based on the assumption that the signal generator delivers random signals, too. If these signals are delivered with a constant frequency, the regular interval length must be much longer than the largest dead time in the system. If this is the case, the equations introduced above can be adapted. Under these conditions the periodic signals get lost proportionally to the random signals, whereas random signals get lost proportionally to the sum of the random and the periodic signals. Therefore, N_{pa} is larger than expected with random signals. As a result, the correction factor f_{td1} of Eq. (4.11) must be changed to

$$f_{td2} = f_{td1} + (N_p/N_a) \cdot (1 - 1/f_{td1})/f_{td1} . \tag{4.14}$$

4.2 Inductive Approach to Uncertainty (Example)

Most people approach uncertainties inductively by investigating the pattern of equivalent data values given in data sets (arrays). The huge advantage of this approach is that nothing must be known about the nature (the characteristics) of the data values except that they are normally distributed (i.e., that the limiting distribution of the relative frequency distribution, see Sect. 5.1, is the normal distribution, see Sect. 5.2.3). One strong disadvantage is that a fairly large number of data values is required. A severe complication can arise when the pattern contains, in addition, some functional dependence.

To ensure that the data set under investigation contains only data values that are random and identically distributed let us use event rates of nuclear radiation of a constant intensity. The results of 12 such measurements conducted under the same experimental conditions over the same lengths of time (1 min each) are listed in Table 4.2, together with the deviations from their mean value y_m (Sect. 4.2.1) and the squares q_i of these deviations.

Actually, repeating a measurement means measuring the time dependence (or time series) of the count rate as given in Table 4.2. In the actual case described here, the true decrease in the count rate is less than 1×10^{-6} between the first and the last measurement (due to a decay time constant of 43 years) that is far beyond the resolution of the numerical value and the precision of the measurement, and can thus be disregarded. The counting loss due to dead time was corrected by applying the live time correction (Sect. 4.1.3).

Although the data in Table 4.2 are sorted with regard to time they are raw data when you are interested in the measured numbers. Usually, one would sort them according to the property of interest by way of a data array to facilitate their analysis (as done in Table 5.1).

Table 4.2. Count rate measurement record

Clock time (h:min:s) t_i	Counts (min^{-1}) y_i	Deviation (min^{-1}) $y_i - y_m$	Squares q_i (min^{-2}) $(y_i - y_m)^2$
12:01:00.00	9975	65	4214
12:02:00.00	9961	51	2593
12:03:00.00	10068	158	24938
12:04:00.00	9805	−105	11043
12:05:00.00	9916	6	35
12:06:00.00	9903	−7	50
12:07:00.00	9918	8	63
12:08:00.00	9882	−28	789
12:09:00.00	9979	69	4750
12:10:00.00	10005	95	9009
12:11:00.00	9708	−202	40838
12:12:00.00	9801	−109	11899
Sum	118921	0	110219
Mean value y_m	9910.1		

Table 4.3. Record of a raw data set

Run number	Clock time (h:min)	Room temperature (°C)	Diode forward voltage (V)
1	8:30	20.0	0.659
2	9:30	21.0	0.657
3	10:30	23.0	0.653
4	11:30	24.0	0.651
5	12:30	24.5	0.650
6	13:30	22.5	0.654
7	14:30	23.0	0.653
8	15:30	22.5	0.654
9	16:30	21.0	0.657

4.2.1 Properties of Data Sets and Arrays

Before we return to our example (Table 4.2), we should understand the difference between raw data and data arrays. An array is an arrangement of raw data into grouped data. Table 4.3 presents a set of three-dimensional raw data, i.e., nine trivariate data points that have been obtained in sequence, as indicated by the run numbers 1 to 9.

Usually, the sequence itself is not relevant if we disregard the fact that sequential means time-sequential. Thus, the sequence numbers are superfluous in this example because they are duplicated by the clock time readings in Table 4.3. These readings indicate that the measurements were done at regular time intervals.

Taking the clock time as independent variable, we get three time series: a two-dimensional one with temperature and forward voltage as dependent variables, and two one-dimensional ones, namely the time dependence of the temperature with a maximum at 12:30 hours and the time dependence of the forward voltage with a minimum of 0.650 V at 12:30 hours.

Of course, one can also investigate the dependence of the clock time or the forward voltage on the temperature or the dependence of the clock time or the temperature on the forward voltage. Obviously, not all of these dependences make sense if a cause–effect relation is considered.

After ordering, e.g., the values of the forward voltage according to their size (from 0.650 V to 0.659 V) these data form a data array. This is usually done for the independent variable.

If many data points are involved it is impossible to understand their collective properties by investigating them individually. Therefore, measures are necessary to describe collective properties of arrays (data sets). These measures are determined inductively from the pattern of the data values.

Measures of Central Tendency

Characterizing a set of data by just one best estimated parameter requires some central property to be found, a *measure of central tendency*. There are several choices:

- The *mean value* y_m (also called the average value or the arithmetic mean) of the n values y_i is defined as

$$y_m = \frac{1}{n} \cdot \sum_{i=1}^{n} y_i \,. \tag{4.15}$$

  The mean is the measure most commonly used. The mean forward voltage of the voltages recorded in Table 4.3 is 0.6542 V.
- The *weighted arithmetic mean* y_{mw} must be used instead if the n values do not have the same precision (unlike in our example). Then the individual values need to be weighted due to their "statistical" quality (Sect. 6.3.1).
- The *median* y_{med} is that value in an array that lies in the center of it, i.e., exactly one half of the data are below it, and one half above it. In cases of even n, the median is the mean of the two values that lie in the center. The median of the voltage readings in Table 4.3 is 0.654 V. Usually the median is very well suited to represent the typical value of the data set because it is, in most cases, less sensitive to singular fluctuations within the data set. This is also true for outliers (Sect. 3.2.7).
- The *mode* y_{mod}, the most frequent value, is the value that has the highest frequency. This should not be confused with the frequency itself with which the mode occurs.

Measures of Dispersion

Measures of dispersion describe the fluctuation of data values within a data set.

- The range

$$R_y = y_{\max} - y_{\min} \tag{4.16}$$

  is the difference between the largest $y_{\max}$ and the smallest $y_{\min}$ data value y_i of the data set. The voltage readings recorded in Table 4.3 have a range of 0.009 V.
- The fluctuation of the individual data values about the mean y_m, the root mean square (r.m.s.) S_m (also standard deviation of the data values) is defined as

$$S_m = \sqrt{\sum_{i=1}^{n} (y_i - y_m)^2 / (n - f)} \,, \tag{4.17}$$

  where n is the number of data points with values y_i. The root mean square is the square root of the quadratic mean of the deviations, i.e., the sum of the squared deviations of all data values from their arithmetic mean divided by the number of *independent* data values, i.e., the degrees of freedom. (As the mean value of the deviations equals zero, following the definition of the arithmetic mean, the quadratic mean of the deviations is the preferred option for a measure of the mean deviation.)

The variable f in the above equation for the root mean square takes into account the Bessel correction. It adjusts the value of the *degrees of freedom* from n to $(n-f)$ to account for *dependences* between the data values and the parameters calculated from them. Here we are dealing with deviations from the best estimate y_m and not from the (unknown) true value. The calculation of the mean value y_m has fixed one of the parameters, the scale, using up one degree of freedom. Therefore, one degree of freedom is lost and the number f has to equal 1. For the voltage data of Table 4.3 we obtain $S_m = 0.0030\,\mathrm{V}$.

Obviously, it is not possible to calculate the standard deviation of a single point. The value of a point is also its mean value, and because $(n - f) = 0$, no independent information remains that could be used to calculate S_m.

- The standard error (or standard deviation) σ of the mean is found by dividing S_m by $\sqrt{n}$:

$$\sigma = \mathrm{S}_m/\sqrt{n}\,. \tag{4.18}$$

  Because the n data values are independent of each other, their uncertainties S_m can be added in quadrature (Sect. 3.4) so that we need to divide by $\sqrt{n}$. The standard deviation σ is the most frequently used measure of dispersion. For the voltage data of Table 4.3 the standard deviation of the mean is $0.0010\,\mathrm{V}$.
- The mean deviation is of no importance in uncertainty analysis. Because the sum of the deviations $\sum(y_i - y_m)$ always equals zero due to the definition of the mean value, this straightforward approach is not suited for characterizing the dispersion. However, if the absolute values of the deviations are used, sometimes a useful measure for the dispersion is obtained: the mean deviation.
- The square of the standard deviation σ^2 is called the variance:

$$\sigma^2 = \sum_{i=1}^{n} \frac{(y_i - y_m)^2}{n \cdot (n-f)}\,. \tag{4.19}$$

  The variance, also called the quadratic mean, is defined as the arithmetic mean value of the square of the deviation of all values from the mean value. Variances often show up in mathematical formalisms, but they have odd dimensions, limiting their usefulness.
- The coefficient of variation $\sigma_r = \sigma/y_m$ (or $\sigma_r = \Delta y_m/y_m$) is used if we are interested in the relative dispersion. This coefficient is dimensionless and is the relative uncertainty (Chap. 1 and Sect. 3.4) of the mean value. Being dimensionless and normalized allows us to compare the dispersion

of quite different types of data. The coefficient of variation of the mean of the voltages recorded in Table 4.3 is 0.15%.

In the case of infinite distributions the variable f is not part of the corresponding theoretical equation because the number of points (and degrees of freedom) is infinite.

Example. *Degrees of Freedom*

During a certain period of time 55 physics students graduated at the University of Vienna, and 8 of these students graduated with honors. At the University of Innsbruck 3 out of 30 physics graduates received their diploma with honors during the same period. Can these two groups of people be statistical samples of the same population?

For the construction of the theoretical population (in the comparison) we need the same total number (85), the same mean percentage of graduates with honors (12.9%), and the same ratio of Viennese students to students from Innsbruck (55:30). Now, when comparing the actual data with the constructed example only one free data point, one degree of freedom remains. Three degrees of freedom are used up for constructing the model, so $f = 3$.

External Uncertainty of Data Values

If data values are random and are identically distributed their random uncertainty can be induced from their dispersion pattern about their best estimate (i.e., their mean value y_m). Obviously, the more values that are available, the better the pattern will be defined. Theory depends on an infinite number of data values. Under these circumstances the theoretical solution is exact.

The fluctuation of the individual data values about the mean y_m, is specified by the r.m.s. value or *standard deviation* of the data values, yielding the probable uncertainty Δy_i:

$$\Delta y_i = S_m = \sqrt{\sum_{i=1}^{n} \frac{q_i}{n-f}} . \tag{4.20}$$

This quantity Δy_i is called *external uncertainty* because it is induced from the external pattern of the data values without paying attention to their individual (internal) properties. *It has the same size for each individual data value.*

Equation (4.20) is based on the assumption (i.e., on the theory) that no changes occur in time. This is not the case for the voltage measurements shown in Table 4.3 because the voltage depends on the temperature, which varies in time. So the voltage data must be corrected for the temperature dependence before the pattern of the data can be used to induce the uncertainty of the voltage measurement.

If there is time dependence the appropriate mathematical function describing this dependence must be used to find the best estimate and its uncertainty. In such cases the deviations from the *individual* best estimates, as given by the function in question, must be used rather than the deviations from a single best estimate, the mean, as done here. If the function is not known, S_m cannot be calculated and no external uncertainties exist. To overcome this, one makes a least-squares *polynomial fit* to the data (Sects. 6.3 and 8.3.3), resulting in fit parameters as the best estimate of the final result.

If not all data values of a set have the same weight (the same precision) no external uncertainties exist, either. Even though it is possible to weight the data points due to their internal uncertainty (see, e.g., Sect. 6.3.1), this leads nowhere: All external uncertainties of one data set must have the same value so that no external uncertainties can exist in such cases. So the precision of such data sets can be described solely by internal uncertainties (see, e.g., Sect. 6.2.5).

The consistency between internal and external uncertainties can be checked with the help of the chi-squared test (Sect. 8.3.3).

Data Reduction of the Example

Count rate data from constant radioactive decay are definitely random and identically distributed. Therefore, the (external) uncertainty can be induced from the pattern of the data. The mean value $y_m = 9910.1$ of the 12 data values in Table 4.2 is the *best estimate* of the number of radiation events in the detector per minute. With $f = 1$ the r.m.s. deviation S_m of all data values becomes $S_m = 100.1$. This quantity S_m is the *external uncertainty* Δy_i of each data value. The external uncertainty of the best estimate is $\sigma = \pm 28.9$, obtained by division of S_m by $\sqrt{12}$. This may be done because these 12 data points are independent of each other so that the law of error propagation (3.13) may be applied. Thus the coefficient of variation is obtained as $\pm 0.29\%$.

Normally one would not attempt to find the cause of the variation within this time series from which the external uncertainty is derived. Most people would just assume that the measurement is the reason for this uncertainty. However, as we will see in Sect. 6.2.1, *this specific uncertainty is not caused at all by the experiment* but is the consequence of the probabilistic nature of the true value. *The process of extracting the external uncertainty of the best estimate from the pattern of individual data values is a kind of inductive inference* (Sect. 8.3.2).

Problems

4.4. Determine, for the following numbers: 4, 5, 8, 9, and 12

(a) the arithmetic mean,
(b) the standard deviation.

4.5. Determine for the following set of data: 4, 4, 4, 4

(a) the mean value,
(b) the median,
(c) the mean deviation,
(d) the standard deviation of the mean value,
(e) its coefficient of variation.

4.6. Determine for the following set of data: 0.87, 1.23, −0.18, −0.55, 0.99, 0.07, 1.77, −1.58, −0.38, −1.23

(a) the mean value,
(b) the median,
(c) the mean deviation,
(d) the standard deviation of the mean value,
(e) its coefficient of variation.

4.7. Determine for the following set of data: 4.90, 11.20, 0.62, 0.38, 6.40, 0.94, 38.20, 0.22, 0.46, 0.24

(a) the mean value,
(b) the median,
(c) the mean deviation,
(d) the standard deviation of the mean value,
(e) its coefficient of variation.

4.8. List

(a) two real cases and
(b) one hypothetical case of data sets for which external uncertainties do not exist.

4.9.
(a) Are all five digits stated for the mean value of the counts recorded in Table 4.2 significant (Sect. 3.2.4 and Chap. 9)?
(b) Present the mean value and its uncertainty with the correct number of digits. (Note: Quoting nonsignificant digits, does, in fact, make sense to have sufficient numerical resolution for the comparison in Sect. 6.2.1.)

4.10. Show the time dependence of the data values of Table 4.2 (by means of a scatter plot, see Sect. 9.1.2), and also the mean value (as a line with a slope of zero). Exclude the origin, and use graph paper or a computer. You may use Fig. 4.3 as a guide.

4.2.2 Reproducibility Within Data Sets

What can we learn from the count rate data in Table 4.2 as gathered by repeating the same measurement? These data are special insofar as their uncertainty (i.e., their standard deviation S_m) is dominated by the statistical nature of radioactive decay. Assuming that the radiation intensity has remained constant over the measuring time and that all data points are of the same precision, the external (uncorrelated) uncertainty of the data values equals the standard deviation S_m. These external uncertainties mirror the *reproducibility* (or *repeatability*) of the measurement result. The reproducibility is also called *precision* of the data and is given by the standard deviation σ of the mean.

The term reproducibility can also be used in connection with other functional dependences between the data. First, one would put a linear dependence to the test using the linear regression (Sect. 4.2.3). In our example, where the measurement time is very short when compared to the half-life of the source, we can approximate the exponential decrease to a first order by a linear dependence.

In Sect. 6.4 we discuss reproducibility in the presence of internal uncertainties.

4.2.3 Linear Regression (Least-Squares Method)

If there are a number of n (≥ 3) bivariate data points with a linear dependence of the variable y on the variable x, one can use linear regression (see also Sect. 6.3, regression analysis) to find the best estimate of the data values, which is a straight line through the data points. The line must fulfill the condition that the sum of the squares of the distances of the points to the line is a minimum (least-squares method). Usually the vertical distances are minimized, i.e., it is supposed that the x-values are exact (i.e., with uncertainties that can be disregarded). After exchanging x and y we still get a line; obviously the equivalent of the above statement is also true for y-values that come "without" uncertainty. In both cases we are dealing with a *one-sided regression*.

The minimal sum of the squares of the deviations is determined by differentiation with the first derivative set to zero. With

$$y = a_1 \cdot x + a_0 \tag{4.21}$$

being the equation of a line, all n pairs (x_i, y_i) must fulfill the following requirement

$$D = \sum_{i=1}^{n} (a_1 x_i + a_0 - y_i)^2 = \text{minimum} . \tag{4.22}$$

The variance σ_y^2 of the y_i (for normally distributed data values) is calculated as

$$\sigma_y^2 = \frac{D}{n} , \tag{4.23}$$

or, taking into account that two degrees of freedom are lost due to the calculation of the two parameters a_0 and a_1, the variance including the Bessel correction is given as

$$\sigma_y^2 = (\Delta y)^2 = \frac{D}{n-2} . \tag{4.24}$$

The parameters a_0 and a_1 (of the line) and their uncertainties Δa_0 and Δa_1 can be calculated the following way (all sums are to be taken from 1 to n)

$$a_0 = \frac{\sum y_i \cdot \sum x_i^2 - \sum x_i \cdot \sum (x_i \cdot y_i)}{n \cdot \sum x_i^2 - \left(\sum x_i\right)^2} , \tag{4.25}$$

$$a_1 = \frac{n \cdot \sum (x_i \cdot y_i) - \sum x_i \cdot \sum y_i}{n \cdot \sum x_i^2 - \left(\sum x_i\right)^2} , \tag{4.26}$$

and

$$(\Delta a_0)^2 = (\Delta y)^2 \cdot \frac{\sum x_i^2}{n \cdot \sum x_i^2 - \left(\sum x_i\right)^2} , \tag{4.27}$$

$$(\Delta a_1)^2 = (\Delta y)^2 \cdot \frac{n}{n \cdot \sum x_i^2 - \left(\sum x_i\right)^2} . \tag{4.28}$$

An interesting property of the regression line is that the y-value corresponding to the mean value x_m of all x_i is y_m, the mean value of all y_i. Therefore, the point (x_m, y_m) lies on the regression line. This can be seen in Fig. 4.3, where the center dashed line has an intersection at the position of the mean value x_m (6.5 min) with the center full line that represents the mean value y_m.

Not only is y a linear function of x, but also x is a linear function of y. Thus, it is quite surprising that, after the application of the one-sided linear regression to both cases, the two answers are not the same! The explanation for this is quite simple: in the first case the squares of the deviations need to be minimized with respect to y, and in the second case this is done with respect to x. In real life the minimization is done with respect to the dependent variable.

A discussion of the results shown in Fig. 4.3 reveals that the fit based on the linear regression suggests that the count rate decreases in time (as expected due to the finite lifetime of radioactive nuclei). This decrease amounts to $(0.14 \pm 0.08)\%$ per counting interval (minute). The decision whether the assumption of a constant count rate or the decrease of the count rate in time, as suggested by the linear regression is "right" (i.e., is the better *best estimate*), will only be possible (in a relevant way) with the knowledge of Sect. 8.3.3. Interestingly enough, both cases are equivalent in one way: For both best estimates the line lies inside the uncertainty bars of 2/3 of the data values, as required.

The regression line is the best linear fit we obtain by minimizing the sum of the squares of the deviations, but the *best estimate* obtained this way is

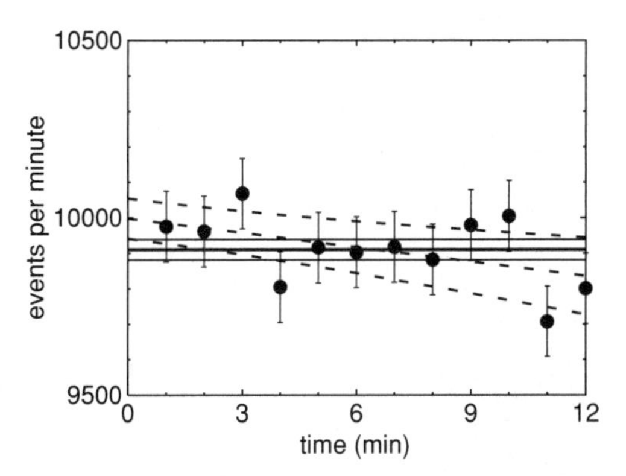

Fig. 4.3. Plot of the data of Table 4.2. The *center dashed line* is the result of the linear regression; *the center horizontal line*, assuming constant radiation intensity, stands for the mean value y_m. The corresponding lines *above* and *below* these alternative best estimates indicate the 1σ confidence intervals

not necessarily the *best approximation* to the *true value* (see the example in Sect. 8.3.3). Note the example in Sect. 10.4.2 of how significantly just one (isolated) data point influences a least-squares result. It should also be clear that even widely differing parameter pairs (a_0, a_1) could deliver very similar data values if the range of interest is very limited. (For illustration: just limit the range to one point, then there would be an infinite number of parameter pairs to represent this point.) Furthermore, it is interesting to see that a_1 is individually important because it describes how changing x influences y (see also Sect. 10.4.2), but for a_0 it is different: a_0 and a_1 are only jointly important.

In Sect. 7.4.2 correlation in connection with linear regressions is discussed.

The linear regression is based on the fact that *all* data values have the same weight, i.e., that they have an *uncorrelated* uncertainty of the same size.

If this is not true, weight factors need to be introduced (Sect. 6.3.2).

Problems

4.11. Show that the equation of the arithmetic mean (4.15) is in accord with the method of least squares.

4.12. A regression line is given by $y = -70. + 3. \times x$. Find the regression line prediction of y_i for $x_i = 55$.

4.13. Carry out the regression calculation for the data in Table 4.2 and show that the y-value of the regression line at the position of the mean value x_m is equal to the mean value y_m of the count rate values.

5

Frequency and Probability Distributions

5.1 Frequency Distribution (Spectrum)

When dealing with a large amount of raw data, often it makes sense first to sort the data according to a specific characteristic and then to analyze the resultant data array. This array can, e.g., be presented as a frequency distribution like the one shown in Fig. 5.1. From this distribution we can tell at first sight that its maximum equals 4, the minimum 1, and that the range is $5 - 1 = 4$. Some scientists call this conversion a transformation of the data values to their frequency space. *In this way the characteristic of the data that is the dependent variable in the data set becomes the independent variable of the frequency distribution.* It is very important to keep that in mind. In Fig. 5.1 the dependent variable of the data set is presented parallel to the y-axis, as usual, but that of the frequency distribution in parallel to the x-axis.

Do not get confused if for that reason the same quantity is named y_i in the data set and x_i in the frequency (probability) distribution.

Figure 5.1 is one of the very few cases where raw data have discrete values in such a way that their size (height) can be used directly as the independent variable of the distribution (spectrum). Usually the quantities are analog, i.e., they must be digitized before they can be grouped into classes (or pulse height channels).

As discussed in Sect. 5.1.1, frequency data are usually grouped into classes in such a way that no gaps appear in the grouped distribution of discontinuous data. Table 5.1 shows just one possibility of assigning classes to the 12 data values of Table 4.2 to yield a continuous frequency distribution.

A simple and frequently quoted example of a frequency distribution is the height of male students measured during routine health checks. Height is an analog quantity that is digitized in the course of the measurement. The maximum number of classes (or categories or bins or channels) depends on the data range and the resolution of the measurement. For instance, if the

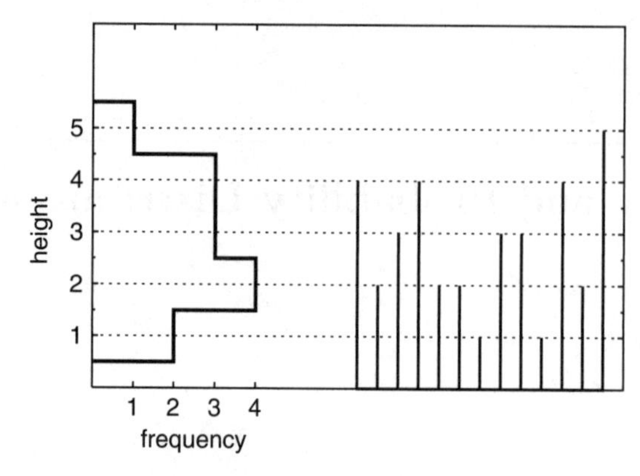

Fig. 5.1. Set of raw data on the *right-hand side*, and their frequency distribution shown as a histogram over the y-axis on the *left-hand side*

Table 5.1. Frequency distribution of the data of Table 4.2

Class value (min^{-1})	Frequency
9730	1
9830	2
9930	7
10030	2

Table 5.2. Frequency distribution of the height of 100 male students

Height (cm)		Number of students
Class limits	Nominal value	
150–155	152.5	0
156–161	158.5	5
162–167	164.5	18
168–173	170.5	42
174–179	176.5	27
180–185	182.5	8
186–191	188.5	0

resolution is 1 cm and the heights lie between 155 and 185 cm, a maximum of 30 classes with the minimum class width (or length or size) and class interval of 1 cm would result.

By class grouping we can reduce the amount of data we deal with. For instance, in Table 5.2 we combined six minimum class widths to get a class interval of 6 cm. This table shows a typical distribution of the height of $n = 100$ students.

The most important characteristic of the classes in Table 5.2 is their nominal value. In the class with a class midpoint of 158.5 cm we find five students

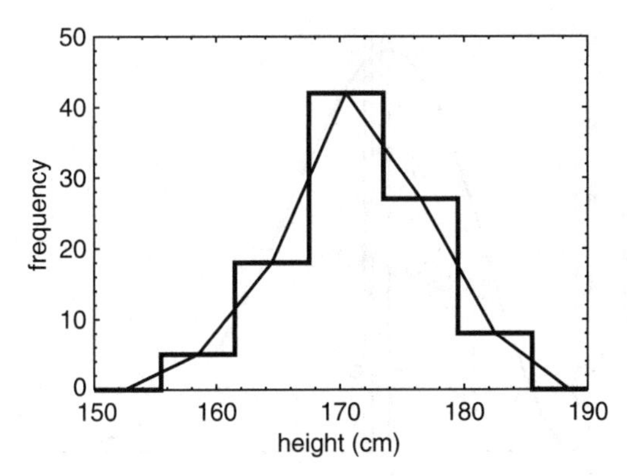

Fig. 5.2. Two common ways of displaying frequency distributions: the histogram and the frequency polygon

with heights between the class limits (or boundaries) 155.5 and 161.5 cm, and in the highest class with a nominal value of 182.5 cm there are eight students with heights between 179.5 and 185.5 cm. To indicate that no students had a height lower or higher than these two extremes, we have added two empty classes. Another way of accomplishing this is introducing open class intervals, i.e., classes with the properties < 161.5 cm and > 179.5 cm.

The mean value $y_m = 171.4$ cm of the height of all 100 students in Table 5.2 is most easily obtained from the frequency distribution by calculating the weighted arithmetic mean (Sect. 6.3.1) using the class frequency as weight.

Such a distribution is best presented with the help of a *histogram*, as in Fig. 5.2. A histogram consists of the outline of bars of equal width and appropriate length next to each other. By connecting the frequency values at the position of the nominal values (the midpoints of the intervals) with straight lines, a *frequency polygon* is obtained. Attaching classes with frequency zero at either end makes the area (the *integral*) under the frequency polygon equal to that under the histogram.

Frequency distributions that are of general interest usually are normalized. This is done by dividing each class number by the total number n; the integral of this new distribution equals one. Distributions constructed this way are called *relative frequency distributions.* If the total number n reaches infinity, relative frequency distributions become *probability distributions.* These are called the *limiting* or *parent distributions* of the relative frequency distributions.

5.1.1 Characteristics of Distributions

The two most important intrinsic characteristics of distributions are mode and symmetry. Depending on the number of equivalent peaks, one speaks of uni-, bi-, tri-, or multimodal distributions. A unimodal distribution is called

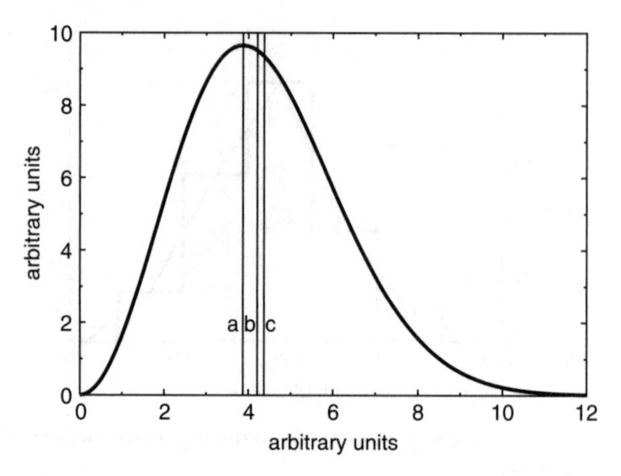

Fig. 5.3. Skewed distribution showing the positions of mode (**a**), median (**b**), and mean value (**c**)

skewed if it is not symmetric with regard to the position of its maximum (to its mode, see Sect. 4.2.1), like the one in Fig. 5.3. In such cases the mode does not coincide with the mean, making an application more difficult. Therefore, in error analysis the asymmetric binomial and Poisson distributions are frequently approximated by the symmetric Gaussian distribution (Sect. 5.2.2).

Measures of Unimodal Distributions

As in the case of data sets, we encounter measures of central tendency and measures of dispersion. Actually, these measures are closely related to those in data sets and consequently are referred to the same way. However, there is a distinct difference in how the measures are determined. The measures of data sets are induced from the pattern of the data values; that is, they reflect *external* properties. Measures of a distribution are intrinsic properties of this particular distribution. They reflect *internal* properties, which can be determined by deduction.

A second difference concerns the presentation of the measures. Measures of raw data sets are connected with the dependent variable (y-values), whereas measures of distributions are connected with the independent variable (x-values). This distinction was discussed in Sect. 5.1.

The measures of central tendency are

- the *mean* $m = x_m$,
- the *median* x_{med}, which splits the distribution in such a way that half of the values are above it and the other half are below it,
- the *mode* x_{mod}, which is the position of the maximum.

For symmetric distributions these three measures coincide. For the skewed distribution in Fig. 5.3 they have quite distinctly different positions.

Aside from the measures of dispersion that we already know, the

- *standard deviation* σ, the
- *variance* σ^2, and the
- *coefficient of variation* $\sigma_r = \sigma/m$, the
- *full-width at half-maximum* (FWHM) is a particular illustrative measure, especially for *symmetric* unimodal distributions. Nevertheless, the standard deviation σ is most commonly used.
- The *range* $R_x = x_{\max} - x_{\min}$ is the difference between the largest $x_{\max}$ and the smallest $x_{\min}$ data value in a frequency distribution. It is of no use in distribution with parent distributions that extend mathematically to infinity. In such cases the size of the range can have nearly any values making this measure meaningless.

In all those cases where the limited number of data values do not allow the determination of the parent distribution, the equations given in Sect. 4.2.1 must be used to induce the value of all these measures. In cases where it is clear which probability distribution applies (as in measurements of radioactive events), the measures are approximately derived from the type of the distribution, even if just one data value of the distribution exists.

5.1.2 Effect of Data Uncertainty on the Distribution

The three main uncertainty components of measured data values are scale uncertainty, interpolation uncertainty, and digitizing uncertainty.

- **Scale Uncertainty.** *The (fractional) scale uncertainty of the measures of central tendency and of the dispersion of the frequency distribution is identical with that of the data values.*
- **Interpolation Uncertainty.** The conversion of analog data to digital data values should be done linearly, i.e., independently of the size. However, some nonlinearity in this conversion is unavoidable, resulting in an interpolation uncertainty. (Ideal interpolation gives ideal linearity!) A spectrum (a frequency distribution) of such digital data values has a class or channel width of one least-significant bit (LSB). Consequently, the class midpoints will not be on a straight conversion line as required for ideal linearity. The largest deviation of the actual conversion (or transfer) curve from this ideal line divided by the full-scale value of the conversion is called the fractional *integral nonlinearity* of the converter. There are two choices of how to construct the nominal conversion line:
  1. use the zero-point and the full-scale-point for the definition (in accord with interpolation)
  2. find a line by the least-squares method (Sect. 4.2.3) best approximating the transfer curve.

  Obviously, the integral nonlinearity distorts the shape of the frequency distribution.

- **Digitizing or quantization uncertainty.** The resolution, the size of the LSB, gives the smallest difference in digitized data values. However, the difference in analog data can be infinitesimal. Therefore, it will happen that two analog data that hardly differ at all will be converted into digital values differing by one LSB. Consequently, this difference is an artifact that is associated with digitizing; it is called digitizing uncertainty. It is basically the same as the rounding uncertainty that we encountered in Sect. 2.3.3. The number 2.5000... is rounded to 3, but 2.4999... is rounded to 2. *The digitizing uncertainty of the data values results in fuzzy class borders of the frequency distribution.*

The Differential Nonlinearity

The differential nonlinearity reflects the variation of the class width within a frequency distribution as a result of the integral nonlinearity. Let us consider a spectrum with the smallest possible class width of one LSB. In the case of an ideal conversion each channel having a width of one LSB would correspond to identically sized intervals of the analog data. Because of the nonlinearity of the conversion, a change by one bit in the digital value corresponds to a different size interval of the analog data, depending on the data value.

Whereas the integral nonlinearity is of importance both for individual data and data arrays, the differential nonlinearity plays a role mainly for spectra (frequency distributions) because differential nonlinearity affects the apparent frequency of the nominal class values. This results in a distortion of the spectrum. To keep this distortion small, the differential nonlinearity must be small. The appropriate specification of the differential nonlinearity is, therefore, given in percent of the mean class (channel) width. It is clear that a specified differential nonlinearity *contributes to the uncertainty* of the frequency of a channel; it must be treated as a *percentage uncertainty* contribution.

When grouping n data channels into a wider class, the effect of the differential nonlinearity decreases with $\sqrt{n}$, assuming that the differential nonlinearity of one channel is independent of the next one, which will be true in many cases. However, when using electronic analog-to-digital converters (ADCs) for the interpolation, this will depend on the principle of the converter.

Examples

1. *Length measurement.* The marking on standard measures is typically done within ± 0.01 cm. Thus, when sorting the height of students into class widths of 1 cm there will be a differential nonlinearity of $\sqrt{2}\%$. Grouping into classes with a width of 6 cm, as done Table 5.2., will then reduce the differential nonlinearity to 0.58%. Therefore, the differential nonlinearity will contribute an uncertainty of about $\pm 0.6\%$ to the uncertainty of each class frequency.

2. *Differential nonlinearity.* In Sect. 4.1 we have seen that the randomness of radioactive events makes the probability of the occurrence of an event within a given time interval length the same, independent of the position of the interval on the time axis.
   Let us assume that 10^4 random signals/s are recorded in a detector and that the time distance of each of these signals from a preceding reference signal (with a frequency of 100 kHz) is measured with a device having a time resolution (channel width) of 10 ns and a range of 1000 channels. After 1000 s the mean frequency in each channel would be 10,000 with a statistical uncertainty of ± 100. The frequency spectrum would be flat with a superposition of a statistical ripple of $\pm 1\%$ (r.m.s., 1σ confidence level).
   If the width of a specific channel would be smaller by 5% (and for compensation some other channel wider by 5%), it would result in a dip of 500 counts below the flat spectrum at this channel (and a peak of 500 counts above the flat spectrum at the other position). Thus, *differential nonlinearity in the data conversion results in a change of the shape of the frequency distribution.*

For some type of converters a 50% differential nonlinearity (expressed flatteringly as 0.5 LSB) is common. Just consider that such a "narrow" channel occurs in the center of a peak. The frequency distribution would show two peaks instead of one! Therefore, the differential nonlinearity of spectral devices should be less than 1% to avoid crass distortions.

Problems

5.1. Convert the frequency distribution of Table 5.2 into a relative frequency distribution.

5.2. Calculate the arithmetic mean value of the 13 pulse heights in Fig. 5.1.

5.3.

(a) Calculate the weighted mean (Sect. 6.3.1) of the frequency distribution of Table 5.1.
(b) Compare this value to the mean value in Table 4.2.
(c) How can this discrepancy be explained?

5.4. Calculate the following measures of the frequency distribution in Table 5.2:

(a) arithmetic mean,
(b) median,
(c) mode,
(d) range,
(e) standard deviation of the mean value,
(f) coefficient of variation.

5.5. For *probability* distributions an empirical relation between mean value, median, and mode is well established: mean – mode $\approx 3 \cdot$ (mean – median). Check whether this also holds true for the following *frequency* distributions, given in

(a) Fig. 5.1,
(b) Tables 4.2 and 5.1,
(c) Table 5.2,
(d) Name possible reasons for the failures.

5.6. Verify the positions a, b, and c in Fig. 5.3.

5.2 Probability Distributions

If the number of data points in a relative frequency distribution is increased to infinity, the distribution becomes a probability distribution. *Statistical predictions due to statistical theory are only valid for probability distributions.* For a finite number n of samples only approximate predictions are possible. For $n \geq 100$ such an approximation should be good enough for all cases. Even though the normal distribution is the usual choice when there is a need of a probability distribution, it is necessary to be familiar with the binomial and Poisson distributions as well.

5.2.1 Binomial Distribution

In the following application random signals (Sect. 4.1) at the output of a detector are counted. A number of $z = 5168$ counts are recorded in $n = 323$ measuring periods of the same length (e.g., 1 s). The probability p of recording counts will be the same for each measurement period, namely $p = 1/323$. Because of the statistical nature of radioactive decay, this does not imply that exactly $x_i = p \cdot z = 16$ counts are recorded in each of the periods. This number $p \cdot z$ is just the *mean value*; the actual number of counts fluctuates around this value. For an infinite number of counts this fluctuation is described via the (theoretical) *asymptotic distribution* of the measured relative frequency distribution.

Note:

- Less than one count is expected to be lost because of the dead time. Therefore no dead time correction was applied.
- Within the data set the count rate is the dependent variable y_i, whereas in the distribution the count rate is the independent variable x_i.

In our case (the number of counts varying randomly in time), the asymmetric probability distribution to use is the binomial distribution P_x:

$$P_x = p^x (1-p)^{z-x} \frac{z!}{x!(z-x)!} \,. \tag{5.1}$$

This distribution is a distribution of discrete values and is defined for positive integers x and z only.

Note : (The symbol ! after z stands for the *factorial*, which means the multiple product of all positive integers that are smaller than z with z. Thus $4! = 1 \cdot 2 \cdot 3 \cdot 4 = 24$; note that $0! = 1$)

The *mean* m of the binomial distribution is given by

$$m = p \cdot z \,. \tag{5.2}$$

Its standard deviation is given by

$$\sigma = \sqrt{m \cdot (1-p)} \,. \tag{5.3}$$

For our example we get $x_m \approx m = 16$ and $\Delta x_m \approx \sigma = 3.994$.

Problems

5.7. Five people A, B, C, D, and E have boxes containing ten identical balls each that are numbered from 1 to 10. Each person (randomly) takes one of the ten balls out of his box. Calculate the probability that at least one of the five balls taken will be numbered 10. (Hint: Use the probability that number 10 is not picked.)

5.8. A marksman is known to hit the bull's-eye with a probability $p = 0.85$. What is the probability of him hitting the bull's-eye $x = 3$ times in the next $z = 4$ shots?

5.9. Use the relation $(\mathrm{n}+1)! = (\mathrm{n}+1) \cdot \mathrm{n}!$ to show that $0!$ should be 1.

5.2.2 Poisson Distribution

The binomial distribution is often approximated by the more convenient Poisson distribution:

$$P_x = (m^x \cdot e^{-m})/x! \,. \tag{5.4}$$

The parameters p and z that occur in the binomial distribution are replaced by their product, the mean value $m = p \cdot z$.

The Poisson distribution can be derived from the binomial distribution for the limits $p \to 0$ and $z \to \infty$. These conditions are fulfilled, e.g., in the case of radioactive sources. Therefore, in such cases the Poisson distribution is used instead of the binomial distribution because it is a one-parameter distribution and is thus easier to handle.

The *mean* of the Poisson distribution is its parameter m; its *standard deviation* is given by

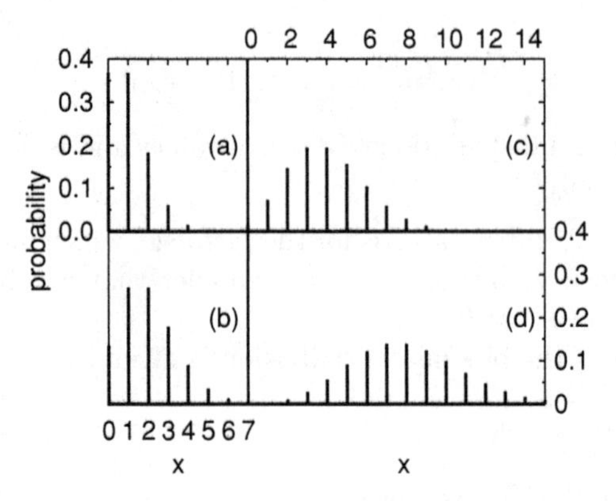

Fig. 5.4. Bar charts of the Poisson distribution for small m for (**a**) $m = 1$, (**b**) $m = 2$, (**c**) $m = 4$, and (**d**) $m = 8$

$$\sigma = \sqrt{m}\,. \tag{5.5}$$

For our example we get $x_m \approx m = 16$ and $\Delta x_m \approx \sigma = \sqrt{m} = 4$.

The recursivity of the Poisson distribution is an interesting feature, especially when writing computer code:

$$P_x = P_{x-1} \cdot m/x\,. \tag{5.6}$$

Like the binomial distribution, the Poisson distribution is also discrete, i.e., both are originally only defined for positive integer numbers x. Consequently, the correct graphic presentation is a histogram (or a bar chart, as shown in Fig. 5.4). For small m both types of distributions are strongly asymmetrical, as can be seen in Fig. 5.4 for the Poisson distribution.

The (theoretical) frequency distribution of the case described above is shown in Fig. 5.5. The frequency, i.e., the number of occurrences of a count rate value x_i, can be read from the y-axis. For instance, the mean value ($x_m = 16$) is recorded 32 times, and the two end values ($x_{\min} = 6$ and $x_{\max} = 27$) occur only once each.

In the chosen case there is no difference between the answers of the binomial and the Poisson distributions. Obviously, both distributions are not symmetric around the mean value, i.e., the median and the mean values do not coincide. As we saw in Fig. 5.4, this feature is even more distinct for smaller mean values m; for larger mean values ($m > 100$) they practically coincide. For such mean values the distribution can be approximated very well by the (symmetric) Gaussian distribution.

According to a rule of thumb, the binomial distribution may be approximated by the Gaussian distribution if $z \cdot p \cdot (1 - p) > 9$, i.e., for $p = 0.5$ one gets $z > 36$, and for $p = 0.1$ one gets $z > 100$.

As the binomial distribution degenerates to the Poisson distribution if z is large and p is small, the corresponding condition for the Poisson distribution becomes $m > 9$.

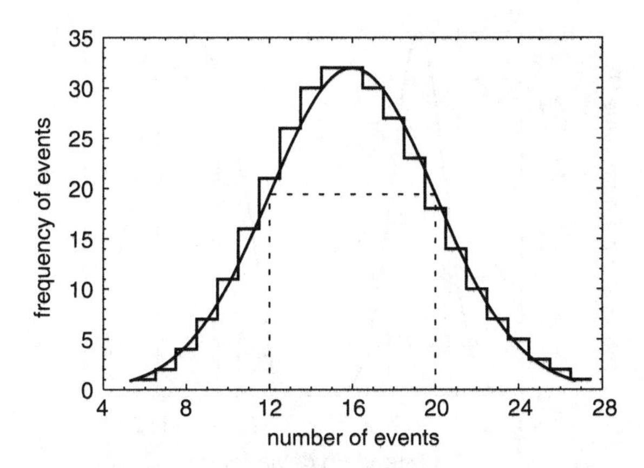

Fig. 5.5. Frequency distribution of the data values x_i. They have a mean value $x_m = 16$ with a standard deviation of $\sigma = \Delta x_m = \pm 4$ and are normalized to a total of 5168 counts. The *smooth curve* shows the Gaussian approximation to this frequency distribution

Problems

5.10. On a meadow with flowers there are, on average, $x_m = 3$ daisies per square meter. Presuming a statistical distribution of seeds, what is the probability of finding exactly $x = 0$ daisies in a randomly chosen square meter? (Remember: $0! = 1$)

5.11. What shape would the frequency distribution of the signals of the example in Sect. 5.2.1 have, if exactly 16 events occurred in each one of the 323 time intervals?

5.12. When bacteria get infected by infectious phages it is expected that the number of phages per bacteria can be described by a Poisson distribution with a mean m given by the ratio of number of phages over that of the bacteria. If you have 1×10^7 bacteria and 2×10^7 infectious phages:

(a) How many bacteria will not be infected?
(b) How many will be infected by at least four phages?

5.2.3 Normal (or Gaussian) Distribution

The (continuous) normal distribution (Fig. 5.6) is frequently used in statistics as it is easy to handle and the vast majority of random processes can be described by it. Unlike the Poisson distribution, the normal distribution is strictly symmetrical.

The normal distribution not being a discrete, but instead a continuous distribution, we have to deal with a continuous probability density function

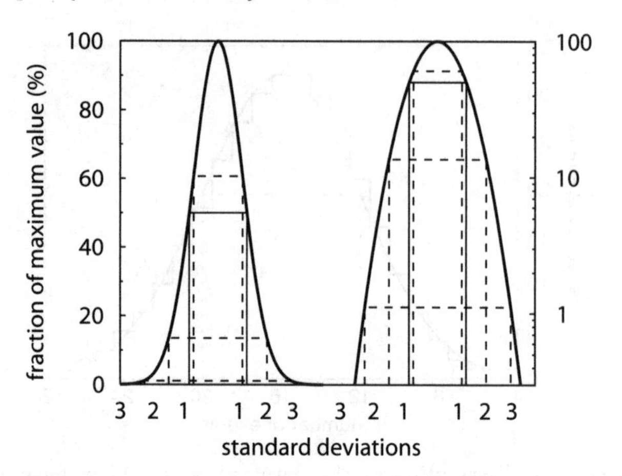

Fig. 5.6. Linear and logarithmic plot of a Gaussian distribution, showing FWHM and the $1\sigma, 2\sigma$, and 3σ confidence intervals

p(x). The probability P that a value x_i occurs between x_1 and x_2 is the area under the curve p(x) calculated between the values x_1 and x_2:

$$P(x_1 < x_i < x_2) = \int_{x_1}^{x_2} \mathrm{p}(y)\mathrm{d}y\,, \tag{5.7}$$

where the Gaussian probability density function $p(x)$ (the normalized normal distribution) is given by

$$p(x) = \frac{1}{\sigma\sqrt{2\pi}} \cdot \mathrm{e}^{\frac{(x-m)^2}{2\sigma^2}}\,. \tag{5.8}$$

Characteristics of the Normal Distribution

The *mean* of the normal distribution is its position parameter m. Because of its symmetry, the *median* and the *mode* also have the same value. The *standard deviation* σ is the width parameter. *For our example*

$$\sigma = \sqrt{m}\,, \tag{5.9}$$

so that one arrives at the same standard deviation $\sigma = \sqrt{m} = 4.0$, as in the case of the Poisson distribution.

The *full-width at half-maximum (FWHM)* of the distribution is very convenient and is

$$FWHM = \sigma \cdot 2 \cdot \sqrt{2 \cdot \ln 2} \approx 2.35\,\sigma\,. \tag{5.10}$$

The *inflection points* of the bell-shaped Gaussian curve are located at the positions $m \pm \sigma$; there the functional value is $\sqrt{\mathrm{e}} \approx 0.6065$ times that at the position m, the maximum. The 1σ *confidence interval* lies symmetrically around the *true value* m with a width of this interval of 2σ or 0.8493 *FWHM*. By integration of the corresponding portion under the Gaussian curve we learn that the probability that the corresponding data values will lie inside this interval is about 68.27% (roughly 2/3); and roughly 1/3 must be outside. Intervals two or three times as large are also used to characterize uncertainty with about 95.45% and with about 99.73% confidence. If not stated differently, the 1σ confidence interval is quoted in physics, giving the *probable uncertainty*.

The probability of a deviation from the mean value decreases strongly with its size (i.e., the distance of the value from the mean value). Very large deviations from the mean value are rare, but possible. Therefore, it is difficult to determine whether a data value that deviates strongly from the others is a legitimate member of the data set or an outlier (Sects. 3.2.7 and 6.4.1).

The Error Function

As discussed before, the probability p of the occurrence of a data value x_i inside an interval between two arbitrary values x_1 and x_2 is given by

$$P(x_1 < x_i < x_2) = \frac{1}{\sigma\sqrt{2\pi}} \cdot \int_{x_1}^{x_2} \mathrm{e}^{-\frac{(y-m)^2}{2\sigma^2}} \, \mathrm{d}y \,. \tag{5.11}$$

After substituting $z = (y - m)/\sigma$ and $\mathrm{d}z = \mathrm{d}y/\sigma$ we get the error function, $\mathrm{erf}(\omega)$, covering the right half of the symmetric curve and normalized to an area of one:

$$\mathrm{erf}(\omega) = \frac{1}{\sqrt{2\pi}} \cdot \int_{0}^{\omega} \mathrm{e}^{-\frac{z^2}{2}} \, \mathrm{d}z \,. \tag{5.12}$$

In error analysis the uncertainty limits $\pm x_2$ are symmetric around the best estimate of the true value m. Therefore one gets, for the probability of the occurrence of a data value inside the uncertainty limits:

$$P(m < x_i < x_2) = \mathrm{erf}((x_2 - m)/\sigma) = \mathrm{erf}(\omega_2) \,. \tag{5.13}$$

Therefore, the independent variable ω of the error function listed in Table 5.3 is given in units of σ. (*Note:* $\omega_2 = (x_2 - m)/\sigma = x_2/\sigma$ because $m = 0$.)

Problems

5.13. With the help of Table 5.3 give the ω-values where the error function has the following values:

(a) 0.6827,
(b) 0.9545,
(c) 0.9973.

Table 5.3. Numerical values of the Gaussian error function. The numbers give the portion of the area under the curve that lies within the limits $\pm\omega$ (in units of the standard deviation σ), thus giving the probability p for a value to lie inside $\pm\omega$

$\pm\omega$	0.00	0.01	0.02	0.03	0.04	0.05	0.06	0.07	0.08	0.09
0.0	0.0000	0.0080	0.0160	0.0239	0.0319	0.0399	0.0478	0.0558	0.0638	0.0717
0.1	0.0797	0.0876	0.0955	0.1034	0.1113	0.1192	0.1271	0.1350	0.1428	0.1507
0.2	0.1585	0.1663	0.1741	0.1819	0.1897	0.1974	0.2051	0.2128	0.2205	0.2282
0.3	0.2358	0.2434	0.2510	0.2586	0.2661	0.2737	0.2812	0.2886	0.2961	0.3035
0.4	0.3108	0.3182	0.3255	0.3328	0.3401	0.3473	0.3545	0.3616	0.3688	0.3759
0.5	0.3829	0.3899	0.3969	0.4039	0.4108	0.4177	0.4245	0.4313	0.4381	0.4448
0.6	0.4515	0.4581	0.4647	0.4713	0.4778	0.4843	0.4907	0.4971	0.5035	0.5098
0.7	0.5161	0.5223	0.5285	0.5346	0.5407	0.5467	0.5527	0.5587	0.5646	0.5705
0.8	0.5763	0.5821	0.5878	0.5935	0.5991	0.6047	0.6102	0.6157	0.6211	0.6265
0.9	0.6319	0.6372	0.6424	0.6476	0.6528	0.6579	0.6629	0.6680	0.6729	0.6778
1.0	0.6827	0.6875	0.6923	0.6970	0.7017	0.7063	0.7109	0.7154	0.7199	0.7243
1.1	0.7287	0.7330	0.7373	0.7415	0.7457	0.7499	0.7540	0.7580	0.7620	0.7660
1.2	0.7699	0.7737	0.7775	0.7813	0.7850	0.7887	0.7923	0.7959	0.7995	0.8029
1.3	0.8064	0.8098	0.8132	0.8165	0.8198	0.8230	0.8262	0.8293	0.8324	0.8355
1.4	0.8385	0.8415	0.8444	0.8473	0.8501	0.8529	0.8557	0.8584	0.8611	0.8638
1.5	0.8664	0.8690	0.8715	0.8740	0.8764	0.8789	0.8812	0.8836	0.8859	0.8882
1.6	0.8904	0.8926	0.8948	0.8969	0.8990	0.9011	0.9031	0.9051	0.9070	0.9090
1.7	0.9109	0.9127	0.9146	0.9164	0.9181	0.9199	0.9216	0.9233	0.9249	0.9265
1.8	0.9281	0.9197	0.9312	0.9328	0.9342	0.9357	0.9371	0.9385	0.9399	0.9412
1.9	0.9426	0.9439	0.9451	0.9464	0.9476	0.9488	0.9500	0.9512	0.9523	0.9534
2.0	0.9545	0.9556	0.9566	0.9576	0.9586	0.9596	0.9606	0.9615	0.9625	0.9634
2.1	0.9643	0.9651	0.9660	0.9668	0.9676	0.9684	0.9692	0.9700	0.9707	0.9715
2.2	0.9722	0.9729	0.9736	0.9743	0.9749	0.9756	0.9762	0.9768	0.9774	0.9780
2.3	0.9786	0.9791	0.9797	0.9802	0.9807	0.9812	0.9817	0.9822	0.9827	0.9832
2.4	0.9836	0.9840	0.9845	0.9849	0.9853	0.9857	0.9861	0.9865	0.9869	0.9872
2.5	0.9876	0.9879	0.9883	0.9886	0.9889	0.9892	0.9895	0.9898	0.9901	0.9904
2.6	0.9907	0.9909	0.9912	0.9915	0.9917	0.9920	0.9922	0.9924	0.9926	0.9929
2.7	0.9931	0.9933	0.9935	0.9937	0.9939	0.9940	0.9942	0.9944	0.9946	0.9947
2.8	0.9949	0.9950	0.9952	0.9953	0.9955	0.9956	0.9958	0.9959	0.9960	0.9961
2.9	0.9963	0.9964	0.9965	0.9966	0.9967	0.9968	0.9969	0.9970	0.9971	0.9972
$\pm\omega$	0.0	0.1	0.2	0.3	0.4	0.5	0.6	0.7	0.8	0.9
3.0	0.99730	0.99806	0.99863	0.99903	0.99933	0.99968	0.99953	0.99978	0.99986	0.99990
4.0	0.9999366									

5.14. Determine the area under the normal distribution between the values 1.53σ and 0.82σ. (Instructions: Use Table 5.3, paying attention to the fact that the area lies only on one side of the maximum.)

5.15. From data obtained by the military from mustering young men, it is known that their height is normally distributed with a mean of $h_m = 176.5$ cm and a standard deviation of $\sigma = 6.0$ cm. In a random sample of 1000 young men, how many are expected to have height:

(a) within the class limits 170.5 cm and 182.5 cm,
(b) in an open class with a lower limit of 180.5 cm,
(c) in an open class with a lower limit of 188.5 cm,
(d) within the class limits of 164.5 cm and 170.5 cm

5.16. The diameter of the central hole in washers is (6.05 ± 0.05) mm, assuming that the uncertainty is the standard deviation of a normal distribution. What are reasons against this? What percentage of the washers is useless if we presume that the washers must be usable for all 6-mm bolts, i.e., for all bolts with a maximum diameter of 6.00 mm (Sect. 3.2.5)?

5.3 Statistical Confidence

Theoretical confidence intervals around the *true value* that are the result of a theoretical probability distribution (assuming an infinite number of points) cannot be applied to results based on a finite number of data points without corrections. The reason is that for an infinite number of data points the uncertainty of the *best estimate* becomes zero, only in which case would it coincide with the *true value.* Thus a finite number of samples decrease the statistical confidence; i.e., the width of a confidence interval must be increased to accommodate the same confidence level.

Probability distributions are based on true values.

The reduction of the statistical confidence due to a limited number n of independent data points is described by the "Student's" distribution. The Student's t factor by which the size of the interval has to be increased depends on the number of degrees of freedom, obviously, but also on the magnitude of the confidence. For instance, when using only six independent data points, 68.27% of the data are situated in an interval of $\pm 1.11\sigma$ (instead of $\pm 1\sigma$), for 95.45% of the data to be inside of this interval its size has to be $\pm 2.57\sigma$ (instead of $\pm 2\sigma$), and to reach a confidence level of 99.73%, the interval has to be $\pm 5.51\sigma$ (instead of $\pm 3\sigma$).

Table 5.4 gives an impression of the increase in interval size necessary to maintain statistical confidence with small numbers of data points.

The corrections of Table 5.4 apply to external uncertainties only, as these uncertainties depend on the knowledge of the true value or the best estimate.

Table 5.4. Effect of the number n of independent data values on the size of the confidence intervals (in units of σ) for three classes of statistical confidence ("Student's" t-factor)

n\Confidence level	68.27%	95.45%	99.73%
3	1.32	4.30	19.22
4	1.20	3.20	9.20
5	1.15	2.78	6.62
6	1.11	2.60	5.51
10	1.06	2.26	4.09
30	1.02	2.05	3.28
100	1.00	2.00	3.04
Ideal	1.00	2.00	3.00

5.4 Dealing With Probabilities

Probability plays a central role in the understanding of uncertainties. Therefore, it is worthwhile to recapitulate some basic probability theory.

- *Probability* furnishes a quantitative description of the likelihood of the occurrence of a specific event. It is usually expressed on a scale from 0 to 1 (or sometimes 0% to 100%) where 0 denotes a completely unlikely event and 1 a certain event.
- Two events A and B are *independent* of each other if the occurrence of one event has no effect whatsoever on the occurrence of the other one; that is, the events do not influence each other. Of course, any number of events may be independent of each other.
- If A and B are independent of each other, then the probability that both of them will occur is the *product* of the two individual probabilities.
- Probability theory rests on the assumption of an infinite number of samples (events). For a finite number of samples, as encountered in reality, probability is approximated by means of the *relative frequency*, just like the true value is approximated by the best estimate. Therefore, predictions derived from probability theory are only approximations for the actual situation (and vice versa).

To promote the ability to handle probabilities some elementary problems and examples are presented.

Problems

5.17. What is the probability p that heads comes on top at the next toss of a fair coin after having had heads three times in a row?

5.18. A fair coin is tossed twice. What is the probability that at least once heads comes on top?

5.19. What is the probability that a die that is not loaded shows an even number when rolled?

5.20. What is the probability that a die that is not loaded shows six at least once after rolling it twice?

5.21. What is the probability that the sum of the points equals 5 after rolling two dice that are not loaded?

5.22. In two consecutive turns a king is drawn from an ideally shuffled deck of 52 playing cards. What is the probability of such an occurrence

(a) if the card is returned to the deck after the first turn and the deck is shuffled anew, or
(b) if the card is not returned?

5.23. In a single draw from a pack of 52 playing cards either a spade or an ace shall be drawn. What is the probability for that? (*Note:* There is one card that is both an ace and a spade!)

5.24. In a deck of playing cards there are, for each of the four suits, two categories of cards: three face cards and ten numbered cards. What is the probability that after randomly picking two cards these two cards belong to the same category *and* one card has a red and the other a black suit?

5.25. A dog is given either one, two, or three treats per day. The probability that he is given one and only one treat equals 0.25, and that he is given exactly two equals 0.50.

(a) What is the probability that he is given three treats?
(b) What is the probability that he is given more than one treat?

5.26. Four streets lead from a crossing following the cardinal directions. It is known that the number of vehicles on these roads is in the proportion 5:3:2:1 (going from north to west). Under these circumstances, what is the probability p_S that a vehicle will head south?

5.27. Of five probabilities the following relations are known: $p_2 = p_1, p_3 = 2 \cdot p_4, p_4 = 2 \cdot p_5, p_2 = 3 \cdot p_5$. Find the probabilities. (Instructions: Search for the smallest probability and use it as reference.)

5.28. It is known that the unemployment rate in a certain country amounts to 8.5% in the rural and 7.0% in the urban population. In this country 60% of the employable population lives in cities. What is the actual average unemployment rate?

5.29. From the laws of genetics it is known that crossing of pure-bred violet-blooming pea plants with white-blooming ones will make $^3/_4$ of the next generation violet-blooming and $^1/_4$ white-blooming. In an experiment where it is not guaranteed that the original plants were purebred we count 131 violet- and 49 white-blooming pea plants. Is this result contradictory to the assumption that the original plants were pure-bred?

5.30. From surveys it is known that (with a certain uncertainty) 60% of the male population buy toothpaste A and 90% of the female population buy toothpaste B. A certain drugstore offers these two toothpaste brands. Toothpaste is to be restocked according to this survey. Name at least one other piece of information necessary for this decision.

Some examples that require more thought:

- In some country two parties A and B run for election. It is sufficiently well known that 60% of the male population favors party A (partly due to the attractive female leading candidate!) and that 60% of the female population favors party B (also partly due to the attractive female leading candidate of party A!). What would the predictions be for the outcome of an election? Without further information about the gender distribution of the voters that can only be known *after* the election – just like the result – it is not really possible to answer this question. It is necessary to form a statistical hypothesis based on, e.g., the gender distribution of the registered voters or the gender distribution of voters in previous elections.
- In medical test series we come across a similar problem. It is (in the best case) only possible to falsify the null hypothesis, i.e., that two types of medication (or some medication and a placebo) have the same effect. How well a certain drug works cannot be shown because a well-founded hypothesis for this would be needed. A success in falsifying the null hypothesis is called *statistically significant* if the probability of a chance result is smaller than 5%, or *highly significant* if it is smaller than 1%, and *of highest significance* if it is smaller than 0.1%.
- A bird watcher spends most of her free time roaming the wilderness looking for rare birds. There is a specific type of bird she has been looking for, but has not observed so far. So she buys a whistle that is claimed to attract this bird with a certainty of 100%. Will this whistle ensure that she will in fact see such a bird?
  Of course, this can only be if this type of bird is present in the area the bird watcher is roaming. Once again: it is not sufficient to know the probability without having relevant information on the population of interest.

These examples should make it fairly obvious that it is not possible to make a (somewhat) valid statement *based on probabilities alone without additional information on the population involved.* Another *example* from the field of medicine shall show us how the use of probabilities can (intuitively) lead to wrong conclusions, and why it is often better to use frequencies instead.

Let us look at a typical medical test used in health checks. In this test there are four types of results:

- positive results that are right,
- positive results that are wrong,
- negative results that are right,
- negative results that are wrong.

In a fictitious examination the probability of a positive result being correct is 95% and that of a negative result being correct is 99.5%. Now, supposing one patient is found to be positive, what is the probability that this patient actually has this illness? As it is 95% of that part of the sample that is positive and 99.5% of that part that is negative further information on the population from which the patients are taken is needed to answer the question.

Health statistics show that for people of this particular age and gender the "illness probability" equals 1:50,000. From 1×10^6 patients of this risk group 19 would correctly be identified as ill, but 5000 patients would get positive results without being ill. Under these circumstances the positive result would only be correct in 0.38% of the cases. Such a test is not suited for examinations in this risk group; it is even irresponsible to use this test that totally unnecessarily suggests further examination in too many cases. If this test were done inside a high-risk group where the illness probability is, for instance, 1:100, two thirds of the positive results would be correct.

Problems

5.31. From a given platform at some train station trains depart in two directions (toward A or B). The departure times of these trains are at regular intervals and always with the same time interval between the trains heading in opposite directions. A retiree who loves traveling (and owns a season ticket for train travel) decides to let chance decide where he will be going by always boarding the train that arrives first. To his surprise he travels to B four times more often than to A (on average). What further information would be necessary for the correct prediction of this result?

5.32. In order to reduce the number of defective goods delivered to the customer it was decided to install a quality assurance machine at the end of the production line. This machine is guaranteed to reject 99.8% of the faulty products. The probability of rejecting flawless products by mistake is 2%. Without use of this machine 3% of the products would be found to be faulty.

(a) What is the probability that a product that has passed will be faulty?
(b) What is the probability that a rejected product will be without faults?

Hint: Use fictitious product numbers, not probabilities.

5.33. A hypothetical, unrealistic problem: At a census in China each person was required to check one of three boxes depending on his (her) belief of how tall a typical American is: 180, 175, or 170 cm. A would-be scientist evaluates exactly 900,000 of these answers. As it happens, the answers are (exactly) evenly distributed among the three choices.

(a) What is the best estimate he gets for the average height of an American? (Being an expert on errors he includes external uncertainties, of course!)
(b) What is basically wrong with his answer?
(c) What is basically wrong with the uncertainty he gives?

6

Deductive Approach to Uncertainty

6.1 Theoretical Situation

In Chap. 5 we found that in the case of random and identically distributed data values (like count rates of radioactive events) the probability of any data value x_i to lie inside the interval $m - \sigma$ and $m + \sigma$ can be predicted if the true value m is known. In this special case for both the Poisson and Gaussian distributions (Sects. 5.2.2 and 5.2.3) the standard deviation is $\sigma = \sqrt{m}$ so that 68% of the data values x_i come to lie inside this interval $\pm\sqrt{m}$, i.e., *any data value* has a probability of 68% to be inside this interval.

Recall discussions in the introductory chapters:

In principle it is impossible to know exactly the true value m.

6.2 Practical Situation

In practice the situation is the other way around: From the data value y (or from all y_i of a data set) a statement about the true value m must be made. From the data value(s) an "estimate" of the true value must be derived. This estimate is called *best estimate* y_m because it is the best approximation of m based on the available data.

Reversing the argument of Sect. 6.1 we get: *The probability that the true value will lie in the interval between* $y_m - \sqrt{m}$ and $y_m + \sqrt{m}$ is 68%. Thus, in the case of a single data value y the best estimate y_m can be given as

$$y_m = y \pm \sqrt{m}\,. \tag{6.1}$$

Obviously, the uncertainty $\sqrt{m}$ depends on the true value m. The true value and therefore also *this internal uncertainty* $\sqrt{m}$ *is a basic characteristic of the proper probability distribution.* Well, the *true value*, the exact value of m, is *unknown in principle*; therefore $\sqrt{m}$ is not known either. Thus we are forced to use $\sqrt{y}$ instead of $\sqrt{m}$ for a single data point, and we get

Table 6.1. Record of the count rate dependence on distance

	Results of experiment		Results of (iterated) fit		
$1/r^2$	y_i	$\pm\sqrt{y_i}$	y_{fi}	y_{mi}	$\pm\sqrt{y_{mi}}$
1.0000	9922	99.6	9959	9962	99.8
0.2500	2539	50.4	2490	2491	49.9
0.0400	405	20.1	398	399	20.0
0.0100	85	9.2	100	100	10.0

$$y_m \approx y \pm \sqrt{y}\,. \tag{6.2}$$

Let us look at a more general case: In Table 6.1 the dependence of the count rate (determined in an experiment, dead time corrected) on the distance r of a (point) source from the detector is recorded. Here the uncertainty of the distance measurement is disregarded. Because of geometrical reasons we expect a dependence on $1/r^2$.

A weighted fit (Sect. 6.3.2) yields better data values y_{fi} (Table 6.1) and uncertainties $\sqrt{y_{fi}}$. Using these improved data values to obtain improved weights for another weighted fit further improves the answer. After multiple iterations (where we always use the latest $\sqrt{y_{fi}}$-value to calculate the weight) we arrive at the best possible of the best estimates y_{mi} of the data values using $\sqrt{y_{mi}}$ for the weight calculation, as done in Table 6.1.

In this special case where the uncorrelated uncertainty primarily depends on the true count rate, using y_{mi} instead of y_i for calculating the weights for the linear regression takes us closer to the "true" value.

Note: This kind of procedure of changing the individual uncertainty values $\sqrt{y_i}$ can, obviously, *not* be applied to data with *external uncertainties* because these are the same for all data points. It is an iterating *weighted* regression.

Above findings are applied in Sect. 6.2.1 to the data values y_i of Table 6.2. Based on the best estimate y_m (the arithmetic mean), the uncertainty of the individual data values y_i should be $\sqrt{y_m}$ rather than $\sqrt{y_i}$

$$\Delta y_i = \sqrt{y_m}\,. \tag{6.3}$$

6.2.1 Best Estimates Using Internal Uncertainties

Negligible Measurement Uncertainties in Nuclear Counting

Table 6.2 presents the data values y_i of the time series of Table 4.2 with their internal uncertainties.

What is the intrinsic reason for the fluctuations of the measured values y_i? As we deal with repeated measurements of radiation with (assumedly) constant intensity, we would expect to get the same result for each measurement. Could it be that the variation is due to measurement uncertainty? This would be in line with believing that measuring is the cause of uncertainty. These count rate measurements are cases of consecutive counting (Sect. 2.1.4) with a (small) uncertainty stemming from the dead time correction. Then there is the measurement of the length of the time interval. As these measurements take place relatively shortly after each other we do not need to worry about long-term time variations of the clock (e.g., due to a change in the ambient temperature). The fluctuations do not depend on the time calibration either because this uncertainty is identical for all intervals. So the absolute calibration of the clock is also not important. Relative variations on the order of 10^{-6} because of uncertainties in the time measurement would be plausible, but the actual fluctuations are about a factor of 10,000 larger.

As discussed in Chap. 4, the reason for the fluctuation is the statistical emission of radiation, a basic characteristic of radioactive decay. So, the fluctuations are not a characteristic of the experiment itself. The *true "value"* is the corresponding binomial (or Poisson) distribution, where the latter is fully described by the *mean value* m and the *standard deviation* σ. The *best estimate* that can be derived from the experimental results consists of the mean value y_m of the data values and its uncertainty Δy_m. The mean value y_m of the 12 (uncorrelated) measurements y_i of Table 6.2 is obtained by dividing the sum by 12, and the uncertainty Δy_m by applying the law of error propagation on the individual uncertainties given in this table. Thus one gets $y_m = (9910.1 \pm 28.7)$ events per minute using internal uncertainties.

The identical result is also obtained without the law of error propagation based on the following thoughts: The total number of detected events in the 12

Table 6.2. Record of count rates with internal uncertainties

Clock time t_i	Measured value y_i	Individual σ $\sqrt{y_i}$	Improved σ $\sqrt{y_m}$
(h:min:s)	(min^{-1})	(min^{-1})	(min^{-1})
12:01:00.00	9975	99.9	99.5
12:02:00.00	9961	99.8	99.5
12:03:00.00	10068	100.3	99.5
12:04:00.00	9805	99.0	99.5
12:05:00.00	9916	99.6	99.5
12:06:00.00	9903	99.5	99.5
12:07:00.00	9918	99.6	99.5
12:08:00.00	9882	99.4	99.5
12:09:00.00	9979	99.9	99.5
12:10:00.00	10005	100.0	99.5
12:11:00.00	9708	98.5	99.5
12:12:00.00	9801	99.0	99.5
Sum	$118,921 \pm 344.8$		
Mean value y_m	9910.1 ± 28.7		

time intervals is 118,921 with an uncertainty of ± 344.8. Dividing by 12 yields $y_m = (9910.1 \pm 28.7)$ events per minute (on average). Since this best estimate is identical with the value obtained above we have shown that the law of error propagation gives the correct answer, at least for this one case. Besides it indicates that in such a situation there is no advantage in performing 12 measurements instead of 1 measurement over the same total time (as far as the precision of the result is concerned).

Problem

6.1.
(a) Determine the mean value y_m of the 12 (uncorrelated) measurements y_i in Table 6.2.
(b) Determine its 1σ uncertainty using the law of error propagation.

6.2.2 Deductive vs. Inductive Uncertainties

At first sight it looks as if an *internal uncertainty* that is deduced from internal characteristics of the data and an *external uncertainty* that is induced from the dispersion of these data values (Sect. 4.2.1) have little in common. If this were so the very good agreement between the internal uncertainty calculated above and the external uncertainty calculated in Sect. 4.2.1 from the data in Table 4.2 ($y_m = 9910.1 \pm 28.9$) would be difficult to explain. (In Sect. 8.3.3 the quality of this agreement is discussed in detail.)

There are two obvious differences to external uncertainties:

- Internal uncertainties exist for single data values.
- Internal uncertainties can differ for each individual data value within a data set. (A purist might claim that these two differences are actually just one.)

Why does it not matter, in above example, whether the external or the internal uncertainties are used? The reason is that our interest lies solely in the true value as approximated by the best estimate. Therefore, uncertainties of individual data values do not show up in the final result (excepting the case of a single data value); only the uncertainty of the true value (i.e., its best approximation) matters. Thus in cases of data sets with data values of equal weights and with a sufficiently large number of data points, internal and external uncertainties will give an equivalent answer. In all other cases internal uncertainties must be used.

In Sect. 3.2.5 we encountered a case where the internal uncertainty (called tolerance) differs markedly from the external uncertainty. Let us revisit this case.

Tolerances of Resistors

Example. ***Resistor Network***

Take resistors with a nominal resistance value of 1 kΩ and a tolerance of 10%. Nowadays, resistors will be produced with very little variation in the resistance values, e.g., ±1%. So if you measure the resistance of a number of 1 kΩ resistors you may get from the measurement pattern a mean value of (1.053 ± 0.013) kΩ which is not in disagreement with the nominal value and its standard deviation obtained from the specification of the producer (1.00 ± 0.10) kΩ. Putting, e.g., 20 such resistors in series a total resistance value of (20.0 ± 0.4) kΩ is expected using the nominal value and the tolerance (the internal uncertainty), but (21.06 ± 0.06) kΩ from the (measured) external uncertainties. The reason for these quite different results is that the resistance values are *not* identically distributed with regard to the tolerance (internal uncertainty). Adding the internal uncertainties linearly (giving the maximum uncertainty, see Sect. 3.2.5) yields (20.0 ± 2.0) kΩ, making this result using internal uncertainties compatible with that using the external uncertainties.

6.2.3 Convolution of Uncertainty Distributions

As shown in Sect. 6.2.1, the same uncertainty is obtained from the sum of all data and from adding the 12 individual uncertainty values in quadrature; this indicates that using the law of error propagation does not change the basic character of the uncertainty distribution. Quadratic addition of Poisson distributions (or Gaussian distributions) results in a distribution of the same kind, but of larger FWHM (according to the law of error propagation). This result can be described as *convolution* of the individual distributions. The convolution of two rectangles of the same width is easily understood: the result is an isosceles triangle with a FWHM of the same length as the width of the rectangles, as shown in Fig. 6.1.

We already know from Sect. 3.4.1 that the standard deviation and consequently the FWHM of a *Gaussian* distribution is only increased very little as a result of the quadratic addition if the FWHM of the second distribution is sufficiently smaller (< one third). For convolutions this means that the (wider) Gaussian distribution remains practically unchanged. Thus it should be plausible that a convolution of a *Gaussian* distribution with any other probability distribution that is a lot narrower than the *Gaussian* distribution leaves the *Gaussian* distribution essentially unchanged.

For the practical error analysis this means that it is sufficient if the *dominant uncertainty components are based on a normal distribution.* If other uncertainty components do not fulfill this criterion, it does not matter as long as their contribution is small enough. The combined uncertainty will then behave as if normally distributed.

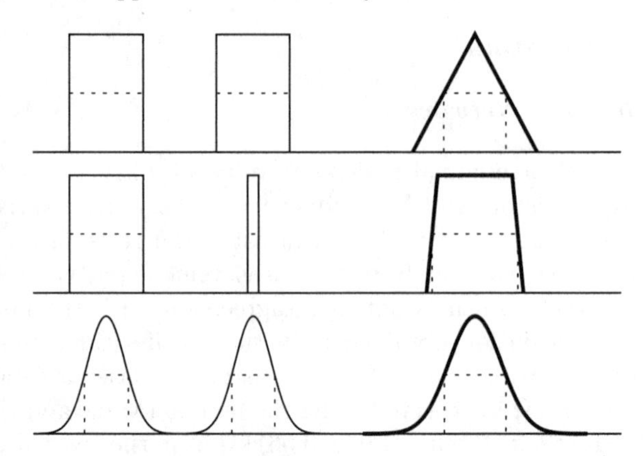

Fig. 6.1. From *top* to *bottom*: Convolution of two rectangular distributions of equal widths, of two rectangular distributions of very different widths, and of two Gaussian distributions of equal width

6.2.4 The Sign of an Uncertainty

As discussed in Sect. 3.2.2, it is a necessary characteristic of an uncertainty that its sign is not known, i.e., that we cannot know whether the *best estimate* is larger or smaller than the *true value*. It is a misconception that the sign of an uncertainty can be determined by repeated measurements. Let us look at the first two measurement values in Table 6.2: The first best estimate amounts to 9975 ± 99.9, the second one to 9961 ± 99.8. The combination of these two measurement values gives a (better) best estimate of 9968 ± 70.6. Obviously this best estimate is smaller than the first value, but larger than the second one. Even though we can determine the deviations of the individual measurements from the combined best estimate this way, we still do not know whether this new best estimate is larger or smaller than the *true value*. Even after the second measurement we still do not know the sign of the uncertainty of the best estimate. The sign of the deviation of an individual measurement from the best estimate derived from it is irrelevant.

We are solely interested in the relation between the true value and the best estimate.

6.2.5 Benefits of Repeated Measurements

Often one encounters a statement such as: "A significant characteristic of random uncertainties is that repeating a measurement 'under identical experimental conditions' not only allows these uncertainties to be better determined, but this also leads to a reduction of the uncertainties". Such a statement appears to be so evident that nobody bothers to substantiate.

For internal uncertainties this statement is strictly false. As discussed further below, there are some benefits in repeating a measurement as far as internal uncertainties are concerned, but there is no uncertainty reduction involved if a fair basis for the comparison is employed (e.g., the same total measuring time in counting experiments, see Sect. 6.2.1 and the example below on nuclear counting).

For external uncertainties this statement is obvious: If you do not repeat a measurement there is no way to induce the external uncertainty from a pattern of the data values. What is the minimum number $N_{\min}$ of data points needed? If you, arbitrarily, limit the uncertainty increase to 10% compared to an infinite number of data points, you need at least $N_{\min} = 5 \cdot f$ data values because of the Bessel correction f, or at least $N_{\min} = 6$ because of the Student's t-factor (for a 1σ confidence value).

However, is repeating really a general method to obtain uncorrelated uncertainties, i.e., can you *always* find "random errors" by repeating a measurement? The example below of a voltage measurement gives a clear answer!

How do we arrive at external uncertainties, at the standard deviation of the data values? We need a function with which we can compare the data. The values of repeatedly measured data are often combined to their arithmetic mean. This means that the assumption was made that the data are time invariant. One uses the function $y = a_0$ with the arithmetic mean y_m being the best estimate of a_0.

Which components of the uncorrelated uncertainty can be obtained by measuring the time dependence? Obviously only those components that fluctuate in time, i.e., those that are uncorrelated in this kind of experiment. In count rate experiments this applies mainly to the effect of the counting statistics, and, maybe, to some effects of fluctuating changes in the environment on the apparatus.

If you want to determine other uncorrelated uncertainty components, e.g., of the angular setting, you must reset the angle for each measurement. Then you will get the superposition of the uncertainty of the counting statistics with that of the readjustments of the angular setting. If you know the time component you may separate the two components by quadratic subtraction (Sect. 3.4.2) because they are obviously independent of each other. *Thus, for each component of the uncorrelated uncertainty one must design a sequence of measurements in which that component is bound to vary randomly.*

Repeating a measurement *under identical conditions* merely gives you a *lower limit* for the uncorrelated external uncertainty because only those uncertainty components are covered that fluctuate in time. Uncorrelated uncertainty components connected with other parameters relevant for your experiment are not covered.

The total uncorrelated uncertainty is usually larger than reproducibility in time (Sect. 4.2.2) suggests, but never smaller.

Designing measuring sequences sensitive to each component of the uncorrelated uncertainty, and determining the resulting standard deviations is tedious, but feasible. How this can be done is shown below under external uncertainties. Another important point (already mentioned in Sect. 4.2.2) is that the determination of the uncertainties of the individual points by means of the standard deviation from their best estimate requires that all points in question have the same (internal) uncertainty. Finally, it is required that the numerical resolution of the data values is sufficient. An insufficient resolution may result in a standard deviation of zero, which obviously does not make sense. In such a case an upper limit (Sect. 3.2.6) may be obtained using the value of the resolution for the uncertainty.

Considering Internal Uncertainties

In Sect. 6.2.1 we saw that the same precision is achieved for a measurement of statistical events if the signals are counted in one single measurement (over the whole measurement time) or if 12 counting periods of the same individual lengths (one twelfth of the total time, each) are used. There is a plausible reason for this: All the information of interest is intrinsic to the aggregation of events. If *identical* events are recorded in the two cases we are comparing, they bear *identical* information.

Still, recording "intermediate results" can be (very) helpful. This allows

1. checking the apparatus, e.g., detecting breakdowns promptly or finding other faults in its behavior,
2. "rescuing" the experiment in case of a breakdown of the apparatus, by using the data recorded up to then,
3. establishing (during the evaluation stage) the foundation of compatibility checks between the experimental results and the theoretical predictions (Sect. 8.3.3).

Considering External Uncertainties

Let us start with an example where uncorrelated uncertainties can *hardly* be determined by time series.

Example. *Voltage Measurement*

Let us return to the example in Sect. 3.1. There we found three types of uncertainties in the case of voltage measurements. What happens with regard to these uncertainties if a given voltage is measured consecutively with the same instrument under identical conditions?

1. *Scale Uncertainty.* The calibration and its uncertainty remain the same for all measurements with this instrument (if no aging or temperature effects occur). Thus we are dealing with a *correlated*, that is, a *systematic*, uncertainty, and consecutive measurements do not enable us to make a statement about it.

2. *Nonlinearity.* From one measurement to the next we will not be able to observe any change in the interpolation because of the great stability of modern electronics. Thus this uncertainty that reflects the nonlinearity remains the same. For this situation the nonlinearity uncertainty is not uncorrelated (random), and consecutive measurements cannot lead to a statement about this uncertainty. Note that this uncertainty is generally uncorrelated for measurements of different voltages as the nonlinearity of the voltage conversion changes from value to value. Nevertheless, this effect cannot be utilized in measurements "under identical conditions".
3. *Digitizing Uncertainty.* This uncertainty occurs twice – for the auto zeroing and for the value of the measured quantity. Instruments with a low resolution will always display the same reading for the same voltage. However, the digitizing uncertainty occurs unavoidably in any type of analog measurements due to the conversion of an analog quantity to a digital value. This uncertainty remains the same for different measurements of the same quantity; it is (systematically) *correlated*. For instruments of a higher sensitivity we detect variations in the display with time. These can be interpreted as different measurement values of the same voltage. They mainly are caused by superposition of electronic noise; its effect is uncorrelated for different measurements. Thus, *only these* contributions to the digitizing uncertainty (and to the total uncertainty of the measurement) can be determined from repeated measurements.

Repeating measurements *under identical conditions* can only be used to determine the external uncertainties of those components of the measurement value that fluctuate in time. The following example (that we will meet again in Sect. 7.3.2 and in Problem 9.1) is intended to demonstrate this point.

Example. ***Influence of Environment***

A highly constant current generator supplies the forward current of 1.0000 mA of a silicon planar diode. The values of the forward voltage, measured every hour in the 1-V range of the same digital multimeter, are listed in Table 6.3.

The mean of the readings is (0.6542 ± 0.0010) V, the mean deviation (the external "uncertainty") of the individual readings is 3.0 mV. How can these findings be explained? The instability of modern electronic instruments (current generator, digital multimeter) will not be observed within a three-digit resolution as used here, as long as the instruments are not exposed to extreme temperatures. So all that can be expected is the fluctuation due to the digitizing uncertainty that, at the utmost, results in a change of ± 1 LSB (least significant bit, here 1 mV). This does not account for fluctuations of up to 5 LSB!

It is known that the forward voltage of silicon diodes changes by -2 mV per degree centigrade. If the nominal forward voltage at 22°C and 1.00 mA forward current is 0.653 V, then it can be concluded that the mean diode temperature during the measurements was (22.6 ± 0.5)°C with individual mean

Table 6.3. Record of repeated voltage measurements

Clock time (h:min)	Readings (V)
8:30	0.659
9:30	0.657
10:30	0.653
11:30	0.651
12:30	0.650
13:30	0.654
14:30	0.653
15:30	0.654
16:30	0.657

variations of 1.5°C. Obviously, the time dependence of the ambient temperature is superimposed on fluctuations that accompany external uncertainties (e.g., the digitizing uncertainty). This temperature effect is not an uncertainty and even less a random uncertainty, but a deviation because of a missing correction (Sect. 7.3.2). Such a correction would require recording the parameter "temperature" at the time of the measurements.

This example demonstrates that the effect of the digitizing uncertainty and that of the temperature mix when measurements are repeated *under identical conditions*. The other uncertainty components that do not change in time do not contribute to the fluctuations and, therefore, cannot be determined by repeating a measurement under identical conditions. However, repeating a measurement "under properly changing conditions" will result in external uncertainties that are a (quadratic) superposition (Sect. 3.4) of the time-connected component with the component for which the changing conditions were designed. They can be separated by subtracting in quadrature (Sect. 3.4.2). Usually, such a procedure is very tedious, and besides it is not always feasible.

6.2.6 Example. *Length Measurement*

The most important three uncertainties of a direct (parallel) length measurement (e.g., by laying down a tape measure) are

- the *scale uncertainty* (which concerns the quality of the relation between the "reference length" on the tape measure and the SI standard),
- the *nonlinearity* (which indicates how evenly the divisions of the reference length were done),
- the *reading* or *digitizing uncertainty* (which shows how well the beginning and the end point of the sample can be transferred to the tape measure and how well their positions can be read, i.e., presented in numbers).

The *random part* of each of these uncertainties can be determined by repeating the measurement "under properly changing conditions".

1. *Digitizing Uncertainty.* The same measurement object is measured several times using the tape measure, and the variation in the readings is recorded.
2. *Nonlinearity.* Again, the same object is measured several times, but using different positions on the tape measure as zero point each time, thus resulting in different end points, too. It is obvious that the answer also contains the random part of the digitizing uncertainty that can be subtracted in quadrature (Sect. 3.4.).
3. *Scale Uncertainty.* For the determination of the random part of the calibration uncertainty via consecutive measurements we would obviously need a different tape measure for each measurement, which makes these experiments quite tedious or at least lengthy. From these experiments we can only gather information about the random part of the calibration uncertainty. Therefore, it is seldom worth the trouble, except if the tape measures have been calibrated independently, so that we can determine the total calibration uncertainty with that procedure. Of course, we have to subtract the uncertainty contributions of all other sources in quadrature, very much as in 2.

Problems

6.2.

(a) Do the uncertainties given for the mean voltage readings and the mean temperature (as read from Table 6.3) provide 68% confidence intervals?
(b) Justify your answer.

6.3. Compare the following quantities of the time series in Tables 4.2 and 6.2, respectively:

(a) the r.m.s. deviation S_m with the internal uncertainty of any individual measurement
(b) the internal with the external uncertainty of the mean value

6.3 Regression Analysis (Least-Squares Method)

It is the aim of all data analysis that a result is given in form of the best estimate of the true value. Only in simple cases is it possible to use the data value itself as result and thus as best estimate. In the example from Sect. 3.2.6 we can clearly see the difference between the data value and the best estimate.

Often the evaluation of data values may be based on theories in such a way that the derived best estimate depends on the data values, but is not identical with them. Corrections might have been applied to the original data values, or the best estimate is an arithmetic combination of data values with its uncertainty calculated by applying the law of error propagation.

So the data values are used to determine the parameters of a mathematical description of a model (a theory), i.e., the parameters of equations. Before determining the best estimate (in form of these parameters) from a data set, it is a good idea to view the data in a graph (Sect. 9.1), typically using a point diagram (scatter plot). This way outliers (Sect. 3.2.7) and isolated points can be identified at first sight. Additionally, it can be seen whether the relation between these data is linear, or if a higher-degree polynomial is needed for their presentation. For the determination of the parameters of a polynomial of the degree g we need (at least) $g+1$ independent data values:

$$y = a_0 + a_1 x^1 + a_2 x^2 + a_3 x^3 + \dots a_g x^g \,. \tag{6.4}$$

For a degree of zero, one constant, i.e., one point, is necessary, for a degree of one (linear equation, a straight line) it is two points, for a quadratic equation (degree=2) three points, etc. Using the logarithm of a simple exponential function yields a straight line (Sect. 9.1.5), therefore two points are sufficient for pinpointing this function.

If the type of functional relation is known, its parameters can be determined via *regression analysis* if at least one free data value is available, i.e., if $n > f$, with n the number of data values and f the Bessel factor (Sect. 4.2.1). This procedure is also called "fitting". Nevertheless, we will reserve the term fitting for those cases where the functional relation is not known, so that the data must be presented by some power series (Sect. 8.3.3).

In regression analysis a curve of the gth degree through a number $n(\geq g{+}2)$ of independent points with the coordinates (x_i, y_i) is arrived at so that the sum of the squares of the deviations of the points from the curve is minimal (*least-squares method*). To this end we need to differentiate the square of the deviations, and the first derivative is then set equal to zero. As can be easily seen, more than $g+1$ points are necessary. If only $g+1$ points were used, the curve would go through all points and thus the sum of the squares of the deviations would automatically be minimal, namely zero. Redundant data values allow determination of the uncertainties of the parameters, and also make it possible to conduct a plausibility test for each data value.

In Sects. 4.2.1 (arithmetic mean) and 4.2.3 (linear regression) we have taken up two easy cases of regression analysis in advance, under the assumption that the uncertainties Δy_i are all of the same size. If this is not the case the precision of the individual data values y_i must be accounted for by their (statistical) weight via the weight factor, w_i, which depends on the internal (uncorrelated) uncertainty Δy_i:

$$w_i = \frac{1}{(\Delta y_i)^2} \,. \tag{6.5}$$

Problem

6.4. Which of the following data sets do not satisfy the conditions for being weights? Find all.

(a) $(w_1 = 1., w_2 = 2., w_3 = 3., w_4 = 4.)$,
(b) $(w_1 = 0., w_2 = 1., w_3 = 2., w_4 = 3.)$,
(c) $(w_1 = 0.01, w_2 = 0.02, w_3 = 0.03, w_4 = 0.04)$,
(d) $(w_1 = -1., w_2 = 1., w_3 = 3., w_4 = 5.)$.

6.3.1 Weighted Mean

If not all n data values have the same precision, the weighted arithmetic mean y_{mw} can be obtained by weighting the individual values y_i with the weight factors w_i:

$$y_{mw} = \frac{\sum_{i=1}^{n} w_i \cdot y_i}{\sum_{i=1}^{n} w_i} . \tag{6.6}$$

Equation (6.6) shows clearly that data values with smaller uncertainties contribute more to the result than those values that are less precise (due to the quadratic weight factors). Similar to the quadratic addition of uncertainties (Sect. 3.4), where we dealt with dominant uncertainties, data values that are more precise by at least a factor of three are dominant. In such a case the other values contribute only insignificantly to the mean value. However, because the precision requirements for data values are higher than for uncertainties, *non-dominant components may not be disregarded!*

A convenient characteristic of the weighted arithmetic mean (just as for the mean) that is especially important for applications is its *recursivity*, i.e., taking the mean of $n-1$ values with the nth value gives the same answer as taking the mean of all n values at once.

The *standard deviation* σ_w of the weighted arithmetic mean (that equals Δy_{mw}) is given by

$$\Delta y_{mw} = \sigma_w = \frac{1}{\sqrt{\sum_{i=1}^{n} w_i}} . \tag{6.7}$$

The weights w_i are also called statistical weights. For dimensionality reasons it should be clear that other types of weights do not qualify in this case. This equation can also be derived by differentiating y_{mw} partially with respect to y_i and then applying the law of error propagation.

Table 6.4. Cumulative counting of random events

Elapsed time (min)	Number of counts
1.000	9,975
4.000	39,811
7.000	69,547
8.000	79,429
9.000	89,408
11.000	109,120
12.000	118,921

Problems

6.5. Derive (6.7) for the case of the weighted mean y_{mw} of two data values. (*Note:* By applying this equation $n-1$ times consecutively, e.g., in a loop of a computer code, the correct answer for the mean value of n data values is obtained.)

6.6. Show that the equations for the weighted mean can be converted to the equations for the unweighted mean (Sect. 4.2.1), assuming equal weights w_i.

6.7. Compute the weighted average of the two numbers 3 and 7 using the weights 0.3 and 0.7.

6.8. Determine the arithmetic mean of the numbers 4, 9, 7, and 3 that occur with a frequency (i.e., with a "weight" of) 4, 3, 2, and 1.

6.9. What important difference is there between nondominant components of uncertainties and nondominant components of weighted means?

6.10. The intermediate readings of a count rate measurement of random signals lasting a total of 12 min are listed in Table 6.4. What is the mean count rate per minute and its uncertainty? (For the graphical answer see Fig. 6.2.)

6.3.2 Weighted Linear Regression

For the linear case (i.e., for curves of the first degree) in Sect. 4.2.3 we looked for a straight line:

$$y = a_1 \cdot x + a_0 \,, \tag{6.8}$$

so that all n data points (x_i, y_i) fulfill the following requirement:

$$D = \sum_{i=1}^{n} (a_1 \cdot x_i + a_0 - y_i)^2 = \text{minimum} \,. \tag{6.9}$$

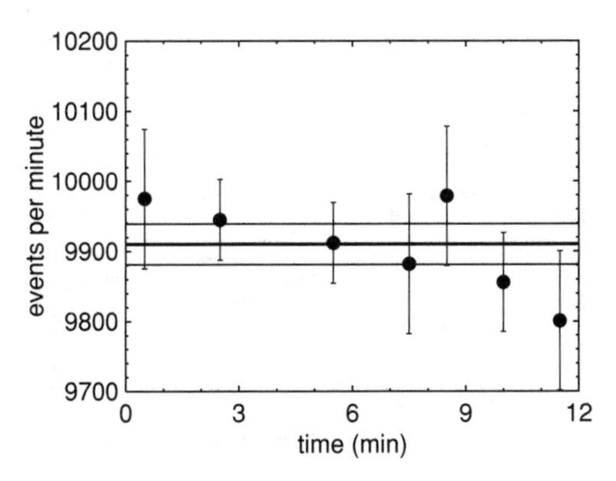

Fig. 6.2. Weighted mean of the count rate data of Table 6.4 together with the 1σ confidence interval

If the data values have *internal* uncertainties Δy_i of different sizes we must assign weight factors w_i according to (6.5).

In addition, we assume that the x_i values have "no" uncertainty (or better: a negligible one). If both the x_i- and the y_i-values have uncertainties that cannot be disregarded, these need to be combined first to a single uncertainty Δy_i (Sect. 9.1.4).

Thus the requirement for the weighted linear regression becomes:

$$D = \sum_{i=1}^{n} w_i \cdot (a_1 \cdot x_i + a_0 - y_i)^2 = \sum_{i=1}^{n} \left(\frac{a_1 \cdot x_i + a_0 - y_i}{\Delta y_i} \right)^2 = \text{minimum} . \quad (6.10)$$

This time it is not the sum of the squares of the deviations that has to be minimized, but the sum of the weighted squares. The close relation to chi-squared (Sect. 8.3.3) can be seen clearly. Weighting the data values with the inverse square of the internal uncertainty gives data values with high precision a much greater influence on the result than those with lower precision (as discussed in Sect. 6.3.1).

The constants a_0 and a_1 and their uncertainties Δa_0 and Δa_1 can be computed using the following equations (with the sums taken from 1 to n in all cases):

$$a_0 = \frac{\sum w_i y_i \cdot \sum w_i x_i^2 - \sum w_i x_i \sum w_i x_i y_i}{\sum w_i \cdot \sum w_i x_i^2 - (\sum w_i x_i)^2} , \quad (6.11)$$

$$a_1 = \frac{\sum w_i \cdot (\sum w_i x_i y_i - \sum w_i x_i \sum w_i y_i)}{\sum w_i \cdot \sum w_i x_i^2 - (\sum w_i x_i)^2} , \quad (6.12)$$

and

$$(\Delta a_0)^2 = \frac{\sum w_i x_i^2}{\sum w_i \cdot \sum w_i x_i^2 - (\sum w_i x_i)^2} , \quad (6.13)$$

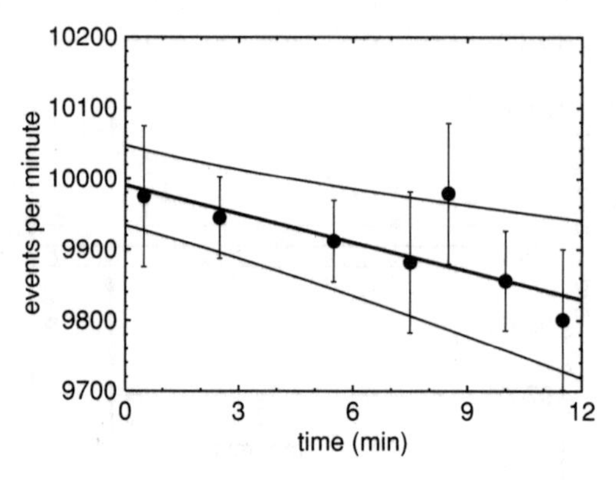

Fig. 6.3. Weighted linear regression applied to the data of Table 6.4. The 1σ confidence interval is also shown

$$(\Delta a_1)^2 = \frac{\sum w_i}{\sum w_i \cdot \sum w_i x_i^2 - (\sum w_i x_i)^2} \,. \tag{6.14}$$

In Problem 6.11 and in Sect. 7.4.1 we give applications of the weighted linear regression.

Problem

6.11. The data of Table 6.4 are known to stem from a radioactive source; i.e., the count rates decrease exponentially in time. Determine the best estimate by way of the parameters of a straight line that best fits the data.
Note: For time intervals that are short compared to the life time of the source, the exponential decay may be substituted by a linear decay. The mean value of the time intervals must be used for the calculation. For the graphical answer see Fig. 6.3.

6.3.3 General Regression Analysis

A linear dependence on two variables can be presented by a plane (two-dimensional linear regression). This case of a multidimensional linear regression can be expressed by equations that are easy to handle. When solving normal equation systems the complexity of the calculations naturally increases strongly with the degree of the polynomial. Nevertheless, it is always possible to find a solution for such polynomials. This is not true for all functions; some normal equations cannot be solved. Such expansive calculations require the help of a computer. Commercial codes are available for doing this job.

Figure 6.3 may assist in understanding the following general characteristics of regression analysis.

1. Multiplication of all data values with a constant k will make all parameters k-times larger. Therefore, the scale *uncertainty* does not affect the regression result. This is the reason why only the data precision (the uncorrelated portion of the uncertainty) enters into the weight factor.
2. Adding a constant d to all data values changes only the constant parameter by d. The (inductive) uncertainty (the pattern of the scattered points) remains unchanged. Therefore, absolute uncertainties must be used for the weight factor rather than, e.g., fractional uncertainties.

6.4 Data Consistency Within Data Sets

Two data values may have the same numerical value, but this does not mean that they are identical. (Identity cannot be shown, only be falsified!) Thus checking the agreement between two or more data values does not mean to check for their equality, but for their consistency, i.e., for their agreement considering their uncertainties.

If we have two independent data values, e.g., of a certain voltage:

- $V_A = (221.1 \pm 2.2)\,\mathrm{V}$
- $V_B = (224.4 \pm 4.5)\,\mathrm{V}$

it is obvious that these two values are consistent because the best estimate of $(221.7 \pm 2.0)\,\mathrm{V}$ as derived from these two values lies inside the uncertainty of either value.

It is possible that one data point (or several points) in a data set deviate notably from the other points. In Sect. 3.2.7 we discussed such outliers. Because outliers have a strong influence on the final result, especially if only few data points exist, it makes sense to search for methods that allow the decision to *discard a data point, even if it is "correct"*, i.e., even if its value is the result of a (rare) statistical fluctuation.

6.4.1 Criterion of Chauvenet

Let us turn to the data in Table 6.5. Basically, it is the same table as Table 4.2, except that the measurement series has been disrupted at 12:06 hours for unknown reasons. The last data value seems to be small. May this value be discarded?

Chauvenet's criterion enables us to check the consistency of a data set using the following simple steps: All data values in a set must have an occurrence probability of at least one, assuming a Gaussian distribution. Under the prerequisite of rounding, we have to use 0.5 instead of 1 for this value. If a smaller probability is calculated for any one data point, it may be discarded, of course, after having taken the precautionary measures discussed in Sect. 3.2.7 ("outliers").

Table 6.5. Disruption of the time series of Table 4.2

Clock time (h:min:s)	Measured value (min^{-1})	Square of deviation (min^{-2})
12:01:00.00	9975	32,701
12:02:00.00	9961	27,822
12:03:00.00	10068	74,966
12:04:00.00	9805	117
12:05:00.00	9916	14,835
12:06:00.00	9040	568,818
Sums	58,765	719,259
Mean value	$(y_{mv} \pm \sigma_v) = 9794.2 \pm 379.3$	
Corr. mean value	$(y_m \pm \sigma_m) = 99445.0 \pm 44.6$	

A suspicious data value y_v in a set of n data values y_i has to undergo the following tests:

- First, the mean value y_{mv} and its standard deviation σ_v are determined.
- Then the probability p_v of the occurrence of a data value that has the same σ-deviation as the suspicious value y_v is computed from the Gaussian error function:

$$p_v = p\left(\frac{|y_{mv} - y_v|}{\sigma_v}\right) . \tag{6.15}$$

This must be true for all n points, so we get $n \cdot p_v > 0.5$ and $p_v > 0.5/n$.

If this is the case, the data value y_v is consistent with the other data values. Table 6.6 lists thresholds for statistically "legitimate" discarding of data, given in units of σ for different numbers n of data points. The Student's correction (Sect. 5.3) has been applied to these data because we are dealing with very few points.

The suspicious value 9040 has a deviation of 754.2 from the mean value. This corresponds to $1.99\sigma_v$ (with $\sigma_v = 379.3$). This distance is greater than the corresponding value of $1.89\sigma_v$ (as given in Table 6.6 for $n = 5$). Therefore this point can be discarded, and we get a best estimate of 9945.0 ± 44.6, instead of 9794.2 ± 379.3. This is a significant disparity – the corrected best estimate is greater by more than 3 (corrected) standard deviations σ. And the deviation of the outlier is now about 17σ instead of the previous $2\sigma_v$.

Before discarding a data point one should investigate the possible reasons for this faulty data value. In the present case a possible reason is quite obvious: The measurement series might have stopped just before 12:06 hours because of a sudden failure of the apparatus, resulting in an effective last time interval of less than 1 min.

Table 6.6. Threshold values for data rejection (in units of σ) below which discarding outliers is statistically not legitimate. The number of data points is n, the table values stem from Table 5.3, and $t(n)$ is the Student's t factor (Sect. 5.3)

n	$0.5/n$	Table value (σ)	$t(n)$	Threshold value (σ)
3	0.167	1.382	1.32	1.82
4	0.125	1.534	1.20	1.84
5	0.100	1.645	1.15	1.89
6	0.0833	1.732	1.11	1.92
8	0.0625	1.862	1.08	2.01
10	0.0500	1.960	1.06	2.08
15	0.0333	2.128	1.04	2.21
20	0.0250	2.242	1.03	2.31
30	0.0167	2.392	1.02	2.44

6.4.2 Discarding Data with Internal Uncertainties

Applying Chauvenet's criterion to the example discussed in Sect. 6.4 is not possible for several reasons. Even if it were possible to find a statistically relevant *discrepancy* between two equivalent data points, which of the two points is "bad"? Furthermore, Chauvenet's criterion is based on external uncertainties, namely on data with uncertainties of the same size.

Consequently, applying it to data with *internal uncertainties of the same size* would easily be possible in a similar way by determining the mean value and its standard deviation. However, it cannot be applied to data with internal uncertainties of different sizes. In practice, though, such data are rather frequent. Thus it would be worthwhile to be able to determine whether an outlier (Sect. 3.2.4) may be discarded for statistical reasons in these cases, too.

There is a way if the number of points available is sufficient for conducting a conclusive chi-squared test (Sect. 8.3.3). You must conduct this test twice: with and without the suspicious data value. If the reduced *chi*-squared that includes the suspicious data value is overly increased, it would be strong evidence that there really is something wrong with this data value.

7

Correlation

7.1 Introduction

The relation between the two characteristics of bivariate data, that is, data having two properties, can be determined via *correlation analysis.* If a statistically relevant, i.e., a probabilistic, relation exists between these properties, we call it a *correlation* between these properties. From the following examples we can see that there are different types of correlation:

- It is possible to determine a simple direct correlation between traffic density and air quality in a city. Of course, the exchange of air is superimposed on this relation, i.e., the wind conditions.
- A simple indirect correlation is the dependence of the need for heating on the latitude. Actually, this need depends primarily on the energy input by the sun. This relation is further modified by the quality of insulation.
- A direct reciprocal correlation, for instance, can be found between the number of children in a family and the percentage of the income that can be spent freely.

Often we can find multiple correlation – effects superimposing on each other – and then there is correlation that can trigger amusement if the correct underlying dependence is not recognized, such as the correlation between income and baldness (age) or between the shoe size and education (age).

Correlation analysis can help us find the size of the formal relation between two properties. An *equidirectional* variation is present if we observe high values of one variable together with high values of the other variable (or low ones combined with low ones). In this case there is a *positive* correlation. If high values are combined with low values and low values with high values, the variation is *counterdirectional*, and the correlation is *negative*.

Problem

7.1. In a mass screening of elementary school children the relation between manual skills and weight has been determined. What other properties of this group of people need to be known to split the data into appropriate classes so that a direct correlation can be determined?

7.1.1 Measure of Relation

If the relation between two properties is to be described, not only the fraction of the corresponding variation must be given, but also the sign of this relation. Furthermore this *measure of relation* has to be independent of the scale chosen; i.e., it must be *normalized* and *dimensionless* to allow analysis of quite different characteristics. It is also necessary that this measure be independent of the size of the data set.

For linear relations the so-called product–moment correlation coefficient r_{xy} (according to Pearson and Bravais) fulfills all these requirements. Its value lies between +1 (for total correlation) and −1 (for total anticorrelation or total negative correlation). It has the value zero if there is no formal linear relation, i.e., if the two properties are independent of each other. The closer the absolute value to one, the closer the linear relation between the two properties under examination. Nevertheless, a correlation close to one is always only a necessary, never a sufficient requirement for a *causal* relation!

The correlation coefficient does not state whether a cause–effect relation actually exists.

With the help of the product–moment correlation coefficient (*see also* Sect. 7.4.2), we can find a linear relation between the x_i and the y_i of the coordinates (x_i, y_i), i.e., of points in a plane. We will try to find out whether these points can be interpreted as points on a line in this plane. Basically, the *linear correlation coefficient* r_{xy} is the sums of the products of the deviations of the values x_i from their mean value x_m with the deviations of the y_i from their mean value y_m, divided by the square root of the products of the squares of the deviations (or the standard deviations σ_x and σ_y) and the degrees of freedom $n - f$:

$$r_{xy} = \frac{1}{n-f} \cdot \sum_{i=1}^{n} \frac{(x_i - x_m) \cdot (y_i - y_m)}{\sigma_x \cdot \sigma_y} . \tag{7.1}$$

The sum of the products in the numerator of the ratio is basically the so-called covariance s_{xy}. By dividing by the two standard deviations we make r_{xy} independent of differences in scale and in the variance of both variables. The method of Pearson and Bravais requires that the data values have a unimodal, symmetrical distribution (e.g., normal distribution), and that the two properties have a linear relation, i.e., their relation is best displayed by a line.

By squaring the correlation coefficient r_{xy} the direction of the correlation is lost, but this way we get the *coefficient of determination* r^2 that allows an illustrative interpretation of the result. A linear correlation coefficient of $r_{xy} = \pm 0.5$ gives a coefficient of determination of $r^2 = 0.25$, meaning that 25% of one variance is determined by the other one (i.e., 25% is correlated and 75% is uncorrelated). The necessity of using r^2 instead of r_{xy} is best demonstrated by the following example:

$r_{xy} = 0.250$ looks like a reasonably strong correlation, but $r^2 = 0.0625$ shows that this is not the case.

7.2 Correlated (Systematic) Internal Uncertainties

Obviously correlation between internal uncertainties can be determined *only* if *more* than one data value is considered, i.e., *it is not possible to determine a correlation for a data value that is based on, e.g., one measurement only.* Or more generally: for a data value with only one property no correlation can exist.

As we will discuss in Sect. 7.2.4 there is no direct way to determine a correlated (i.e., systematic) uncertainty by induction. However, the deductive approach allows us to determine both the uncorrelated and the correlated components of an uncertainty directly. There is no need to find the *degree of correlation* k (Chap. 8) between internal uncertainty components of empirical data. We use the two possible extremes only:

1. *uncorrelated uncertainties* (or, if phrased less precisely, *random* or *irregular "errors"*) with the degree of correlation of $k = 0$,
2. *(totally) correlated uncertainties* (or *systematic* or *regular "errors"*) with a degree of correlation of $k = 1$.

So far, we have almost exclusively been dealing with *uncorrelated* uncertainties. The classic example of such uncertainties is the result of count rate experiments based on radioactive decay (as discussed in Chap. 4). The following sections will be dedicated to *correlated* uncertainties.

Totally correlated uncertainty components are *identical*, i.e., they are identical in size and sign (which *necessarily* remains *unknown*). It is a *necessary*, but not necessarily a *sufficient* requirement for a total correlation between two uncertainties that these can be represented by the same number.

7.2.1 Sign of a "Systematic" Uncertainty

The sign of an uncertainty is in principal *always* unknown – as has been stated several times. It is a wrong supposition commonly stated that this is not true for "systematic errors". Let us examine the following example.

Example. ***Voltage Measurement***

Two voltmeters A and B were calibrated (Sect. 7.2.3) independently using a voltage of 10.0000 V, namely within ±0.5% for instrument A and within ±0.1% for instrument B. These calibration processes are independent of each other and consequently the uncertainties are uncorrelated. The (simultaneous) measurement of a given voltage would result in $V_A = (9.950 \pm 0.050)$ V for instrument A, and $V_B = (9.995 \pm 0.010)$ V for instrument B. These uncertainties contain only the scale uncertainties because the voltage chosen is very close to the calibration voltage, thus the digitizing and interpolation uncertainties (nonlinearity) are minimized (or negligible, respectively) because of quadratic "error" addition. Therefore we get a combined best estimate of $V = (9.9933 \pm 0.0098)$ V because the calibration has been performed independently. It is not known how this value deviates from the true value (i.e., the direction of this deviation cannot be determined), therefore the sign of the uncertainty of the best estimate remains unknown. However, a new calibration of both instruments based on this measurement is possible. Measurements using instrument A can be corrected using the factor 1.0044, and those conducted with instrument B using a factor of 0.9998. After applying this correction factor the scale uncertainty of *each of the two* instruments is reduced to ±0.098%. However, these scale uncertainties are correlated.

As indicated in Sect. 3.2.2, the following interpretation is also possible: as soon as a deviation (and its sign) becomes known, it is necessary to correct the original value until no residual deviation can be determined. The sign of the "remaining" uncertainty (±0.098% in our case) is still unknown.

7.2.2 Differentiation From Uncorrelated Uncertainties

The fact that the *same* uncertainty (e.g., scale uncertainty) is uncorrelated if we are dealing with only one measurement, but correlated (i.e., systematic) if we look at more than one measurement using the *same* instrument shows that *both types of uncertainties are of the same nature.* Of course, an uncertainty keeps its characteristics (e.g., Poisson distributed), independent of the fact whether it occurs only once or more often.

Because of the great confusion in the literature (in books and scripts), where the use of the term "error" leads to wrong conclusions, it appears necessary to state the *difference between correlated and uncorrelated uncertainties* once more.

The *sole* difference between correlated and uncorrelated uncertainties is that in the case of correlated uncertainties the uncertainty components are dependent on each other, and in the other case they are *not.*

Uncertainty of a Single Data Value

By now it should be clear that

1. a single scientific data value must have an uncertainty,
2. a single data value cannot have an external uncertainty.

Therefore, a single data value must have an internal uncertainty. But is this uncertainty uncorrelated? Of course, because there is no partner for a correlation. This is true even for the scale uncertainty (e.g., of an instrument), which usually is thought to be systematic (i.e., correlated).

A similar situation exists for any mean value. Although it is derived from a number of individual data values, it actually is a single data value, the best estimate of the data set. If done correctly it will have a (combined) uncorrelated uncertainty component and a combined correlated component. For this single mean value these two components can only be uncorrelated to (respectively independent of) each other; therefore they may be added in quadrature (Sect. 3.4). In Sect. 8.2 we will discuss when it is advisable to keep these two uncertainty components apart.

However, it is possible for a single data value to have a correlation between its uncertainty components (previously mentioned in Sect. 3.4). To make this absolutely clear we will consider the following example.

Example. *Transmission Line*

The delay time of a coaxial cable shall be determined via the signal speed in it. The following experimental setup is used for this purpose: 10 m of the coaxial cable is stretched between two clamps on either end, one end being the electrical input, and the other the output.

Straight to the input an electrically biased avalanche photo diode is connected; the output of the cable is short-circuited. In addition, there is a single pulse laser available that illuminates, with the help of a beam splitter, both the photo diode at the input and a mirror situated at the output of the cable. Also the mirror is so oriented that the reflected laser beam hits the photo diode. With much care the optical length and the cable length are matched (as well as possible).

A fast digital oscilloscope connected to the input is used as measuring device for three marks on the time axis:

- t_1, the moment when the rising slope of the electrical signal triggered by the laser pulse occurs;
- t_2, ditto for the reflected laser pulse;
- t_3, the moment when the falling slope of the signal reflected at the end of the cable reaches the input of the oscilloscope.

The signal speed v in the cable is given by

$v = l/t\,,$
$t = 0.5 \cdot (t_3 - t_1) = 0.5 \cdot t_c\,,$ and
$l = \mathrm{c} \cdot 0.5 \cdot (t_2 - t_1) = \mathrm{c} \cdot 0.5 \cdot t_l\,,$ with
$\mathrm{c} = 299{,}792{,}458\,\mathrm{m/s}$, the speed of light in vacuum (Sect. 2.1). So we get:
$v = (\mathrm{c} \cdot 0.5 \cdot t_l)/(0.5 \cdot t_c) = \mathrm{c} \cdot t_l/t_c\,.$

Thus, both the length of the cable and the travel time of the signal in the cable are measured via time interval measurements with the same instrument, the oscilloscope. Therefore, their uncertainties are highly correlated. So we have one data value, namely the signal speed in the cable but correlated uncertainty components. In Sect. 8.1.4 we will learn how to deal with them.

If a high accuracy is required some additional information is necessary:

- air pressure to correct the speed of light,
- ambient temperature to correct the speed of light, and to be used as a parameter for the cable properties,
- uncertainty in the time readout of the oscilloscope and its nonlinearity,
- uncertainty in the matching process of optical length and cable length.

For air with a pressure of 0.973 atm (= 96.0 kPascal) and a temperature of 22°C the speed of light (at the laser frequency used) is reduced by a factor of 1.00030. The uncertainties of the time measurements are: calibration uncertainty ±0.1%, interpolation uncertainty given as full-scale value divided by twice the memory depth (in our case ±0.0015 ns), and digitizing uncertainty ±0.06 ns. According to the specifications of the instrument one should add the uncertainty components linearly to cover the worst-case situation (Sect. 3.2.5). Not expecting correlation among these components, we will stick to the probable uncertainty and will add in quadrature. The matching of the optical length and the electrical (= cable) length can be done with some care within ±0.001 m.

The measurements yield: $t_l = 66.726\,\mathrm{ns}$ and $t_c = 99.843\,\mathrm{ns}$. Not considering the matching uncertainty, we get

$$v = c' \cdot t_l/t_c = (0.66811 \pm 0.00072) \cdot \mathrm{c}$$

with c' the actual speed of light in air. Using the intermediate results of

$$l = 0.5 \cdot c' \cdot t_l = 0.5 \cdot (19.998 \pm 0.027)\,\mathrm{m} \quad \text{and}$$
$$t = 0.5 \cdot t_c = 0.5 \cdot (99.843 \pm 0.117)\,\mathrm{ns}$$

one gets

$$v = (0.6681 \pm 0.0012) \cdot \mathrm{c}\,.$$

In this case the uncertainty is considerably higher because the correlation of the calibration uncertainty was wrongly disregarded. The explicit determination of l shows that the matching uncertainty may remain unconsidered because the dominant uncertainty is larger by more than a factor of 10 (respectively of 5, if the scale uncertainty is not included in the length uncertainty, as it should be).

Thus the consideration of correlation between the uncertainty components resulted in a total uncertainty that is nearly a factor of 2 smaller.

Uncertainties in Multiple Measurements

Example. *Gas Container*

If a gas-filled container is used for an experiment, the uncertainty of the pressure measurement is 100% correlated for all measurements where the amount and type of the gas inside the container have remained the same. If another pressure (or a different type of gas) is used, the uncertainties are no longer 100% correlated. On the other hand, to get uncertainties without any correlation it would be required that a different, independently calibrated pressure gauge be used for each measurement.

Example. *Background Radiation*

An even more obvious example is the measurement of some radioactive background. Its uncertainty is *always random*, since it is dominated by *counting statistics*. If the same background measurement is used in connection with several foreground measurements (using the same experimental setup, of course), the uncertainty of the background is 100% correlated (systematic) for these measurements.

The ratio of the source strengths of two weak radioactive sources should be measured. For this purpose the radiation emitted from these sources is measured under otherwise identical conditions with the same detector. To correct for the unavoidable radiation background, a background measurement is done, too. To simplify the procedure, the three measurements are performed for 10 min each (live time, see Sect. 4.1.3), yielding the following count numbers: foreground$_1$ $N_{f1} = 1827$, background $N_b = 573$, foreground$_2$ $N_{f2} = 3081$. (*Note:* The count rate is so low that no dead-time correction is necessary.)

The ratio R_s of the source strengths is given by

$$R_s = \frac{k \cdot (N_{f1} - N_b)}{k \cdot (N_{f2} - N_b)} = \frac{N_{f1} - N_b}{N_{f2} - N_b} \,. \tag{7.2}$$

With

$$\partial R_s / \partial N_{f1} = 1/(N_{f2} - N_b) \,, \tag{7.3}$$

$$\partial R_s / \partial N_b = (N_{f1} - N_{f2})/(N_{f2} - N_b)^2 \,, \quad \text{and} \tag{7.4}$$

$$\partial R_s / \partial N_{f2} = -(N_{f1} - N_b)/(N_{f2} - N_b)^2 \,. \tag{7.5}$$

One gets by means of the law of error propagation

$$\frac{\Delta R_s}{R_s} = \pm \frac{1}{(N_{f1} - N_b) \cdot (N_{f2} - N_b)} \times \tag{7.6}$$
$$\sqrt{(N_{f2} - N_b)^2 \cdot (\Delta N_{f1})^2 + (N_{f1} - N_{f2})^2 \cdot (\Delta N_b)^2 + (N_{f1} - N_b)^2 \cdot (\Delta N_{f2})^2} .$$

With $\Delta N_x = \pm\sqrt{N_x}$ the ratio becomes $R_s = 2.000 \pm 0.083$.

Using the intermediate results $(N_{f1} - N_b) = 1254 \pm 49$ and $(N_{f2} - N_b) = 2508 \pm 60$ the ratio would be $R_s = 2.000 \pm 0.092$. This result is wrong because N_b (and the uncertainty of N_b) is identical in both results, so that this correlation between the intermediate data values must not be disregarded.

This simple example demonstrates:

1. Random uncertainties can be systematic.
2. Step-by-step calculations of uncertainties should be avoided, i.e., intermediate results should not be used for the determination of the total uncertainty without taking possible correlation into account.

Note: The counted events of the background measurement are random and independent of each other, i.e., uncorrelated (due to the origin of background radiation), so that their deviations from the mean value in a measurement series can be used to determine the external uncertainty (Sect. 6.2.5). In spite of that the values (and the uncertainties) of the background N_b as used for these two radiation measurements are totally correlated because they are identical. Thus, depending on the circumstances the same uncertainty (ΔN_b) can be *uncorrelated* (in a single application) or *correlated* (in the application shown above). Again we see that both types of uncertainties are of the same nature.

Example. ***Voltage Measurement***

A voltage is to be measured in the 10-V and in the 100-V range of the same voltmeter. The readings are $V_1 = 10.000$ V and $V_2 = 10.00$ V. The scale uncertainty (of the calibration at 10.000 V) is ±0.01%, the interpolation uncertainty equals ±0.01% of the range value, and the digitizing uncertainty amounts to ±1 unit in the last digit (1 LSB).

(a) Determine the total uncertainty of the 10.000 V voltage measurement (in the 10-V range).

Solution: The reading gives $V_1 = 10.000$ V. The calibration has been done for 10.000 V, therefore no interpolation has been done, and the corresponding uncertainty equals zero. The digitizing uncertainty of 1 LSB is ±0.001 V, and the scale uncertainty is ±0.01% of the measurement value, namely ±0.001 V. Therefore the total uncertainty is

$$\Delta V_1 = \pm\sqrt{0.001^2 + 0.001^2}\,\mathrm{V} = \pm 0.0014[14]\,\mathrm{V}.$$

Note: All uncertainties are independent of each other, thus they are uncorrelated and must be added quadratically. The digits given in parentheses would not be relevant in a final result, as they are nonsignificant; for intermediate results (and comparisons) the quotation of these additional digits is legitimate, even necessary.

(b) Determine the total uncertainty of the 10.00 V voltage measurement (in the 100-V range).

Solution: The reading is $V_2 = 10.00\,\mathrm{V}$, the scale uncertainty is $\pm 0.01\%$ of the value, which is $\pm 0.001\,\mathrm{V}$, the interpolation uncertainty equals $\pm 0.01\%$ of the range value, namely $\pm 0.01\,\mathrm{V}$, and the digitizing uncertainty is one unit of the last digit: $\pm 0.01\,\mathrm{V}$. These uncertainties can be added quadratically because they are independent of each other (there is no correlation; see Sect. 3.4), and we get

$$\Delta V_2 = \pm\sqrt{0.001^2 + 0.01^2 + 0.01^2}\,\mathrm{V} = \pm 0.014[18]\,\mathrm{V}.$$

(c) Combine the two measurement values to one result and state its uncertainty.

Solution: The combined result V is obtained via the weighted mean (6.6):

$$V = \frac{w_1 \cdot V_1 + w_2 \cdot V_2}{w_1 + w_2}, \tag{7.7}$$

where the weights w_i are

$$w_i = \frac{1}{(\Delta V_i')^2}. \tag{7.8}$$

The uncertainties $\Delta V_i'$ must be uncorrelated uncertainties. The scale uncertainties stemming from the calibration of the same instrument can be assumed identical for both cases and are therefore correlated. From $V_1 = V_2$ one gets $V = 10.000\,\mathrm{V}$. The standard deviation σ_w of the weighted mean gives the total uncorrelated uncertainty (6.7):

$$\sigma_w = \frac{1}{\sqrt{\sum_{i=1}^{n} w_i}}, \tag{7.9}$$

and

$$\Delta V' = \sigma_w = \pm\frac{\Delta V_1' \cdot \Delta V_2'}{\sqrt{\Delta V_1'^2 + \Delta V_2'^2}}. \tag{7.10}$$

From $\Delta V_1' = \pm 0.001\,\mathrm{V}$ and $\Delta V_2' = \pm\sqrt{0.01^2 + 0.01^2}\,\mathrm{V} = \pm 0.014\,\mathrm{V}$ the total uncorrelated uncertainty becomes $\Delta V' = \pm 0.000997\,\mathrm{V}$.

Since the total uncorrelated uncertainty is independent of the scale uncertainty, these two can be combined quadratically (Sects. 3.4 and 8.2.1) to get the total uncertainty $\Delta V = \pm 0.0014[12]\,\mathrm{V}$.

This value is less than ΔV_1 or ΔV_2, as it should be. On the other hand, it is so close to $\Delta V_1 = \pm 0.0014[14]\,\mathrm{V}$ that it is easily seen that the second measurement does not really improve the total accuracy. In practice it would not make sense to conduct this additional measurement in the 100-V range.

Problems

7.2. The following voltages of a voltage divider are measured using a digital voltmeter in the 1-V range (scale uncertainty ±0.02%, i.e., the uncertainty equals ±0.02% of the reading; other uncertainties, like nonlinearity, ±0.02% of the measurement range): $V_1 = 1.00000\,\mathrm{V}$, $V_2 = 0.90000\,\mathrm{V}$. This results in a measured voltage attenuation of $A_m = 0.90000$.

Determine the uncertainty of such a voltage attenuation for the following two cases:

(a) V_1 and V_2 were measured (consecutively) with the same instrument
(b) V_1 and V_2 were measured "simultaneously" with two different instruments – assuming that the production and calibration of these two instruments happened *entirely* independently of each other.

Notes:

- When in doubt see Sect. 8.1.4 for calculations with correlated uncertainties.
- We are interested in the ratio of the two measurements and not in the transfer characteristic of the (unloaded) voltage divider.
- The instruments have been calibrated at 1.00000 V in the 1-V range. Thus the nonlinearity uncertainty can be disregarded for measurements of 1.00000 V.

7.3. An insecure experimenter decides to avoid correlation at all. So he splits (in the background radiation example of Sect. 7.2.2) the measuring time of the background measurement in two, yielding an intermediate readout $N_{b1} = 276$ of the background counts after 5 min.

(a) Determine the ratio of the source strengths (with uncertainties, of course!) under these changed conditions.
(b) Why is the accuracy of the result not much different?

7.4. Verify the numerical results of the example *transmission line*.

7.2.3 More Examples of Correlated Uncertainties

Calibrating an Instrument

Investigating the calibration process may lead to a better understanding of the scale uncertainty of instruments. In the simplest case the calibration uncertainty consists of the following two components:

1. the uncertainty of the calibration standard,
2. the uncertainty in adjusting the reading of the instrument to be calibrated to the nominal value of the standard when this standard is measured.

Therefore, the calibration process is just an (individual) measurement with an uncertainty that, in general, is quite independent of any preceding calibration process. Thus the two extreme cases when measuring with two instruments of the same kind are:

- The instruments are of the same type (or from the same factory): It can be expected that the *same* standard was used for the calibration, therefore the uncertainty of the standard can be assumed to be 100% correlated. If, on the other hand, the calibration process has been undertaken individually, the corresponding uncertainties are (mostly) uncorrelated.
- If the instruments were produced by different companies (in different countries), the uncertainty of the calibration standard is not necessarily strongly correlated; it may even be assumed to be uncorrelated.

Example. ***Correlated Measurements***

In some laboratory two instruments with an accuracy of $\pm 1\%$ are available for conducting a certain experiment. For now, let us assume that this stated uncertainty is a 1σ uncertainty. In general, companies do not make any statement about this. In most cases assuming a 3σ uncertainty will be closer to the truth.

If only one measurement is done, only one instrument is used, and the question of a correlation does not arise.

1. If two or more measurements are done with one instrument, the calibration uncertainties of all these measurements are *identical. (The deviation from the true value is the same, but still entirely unknown, both in size and sign!).* The scale uncertainty of all these measurements is totally correlated (systematic).
2. If two or more measurements are done using both instruments, *not* all measurements have the same scale uncertainty, even if (as in our example) they should have the same numerical value, and the calibration uncertainties are *not* (totally) correlated (i.e., they are not systematic).

Cross-Section Measurements

When neutrons are produced via the reaction $^{1}\mathrm{H(t,n)}^{3}\mathrm{He}$ by shooting triton beams of *various* energies at a hydrogen gas target the differential cross sections of this reaction can be measured at *various* angles with the help of a neutron detector. Which of the uncertainties encountered in such experiments are correlated (systematic) and which are not?

Discussion

Note: The discussion that follows is somewhat exaggerated; some parts would just be impractical. However, this seems to be necessary for enhancing some important aspects.

(1) General Properties

Energy and Angle Uncertainty. We need to differentiate between two quite different measurement tasks:

- For measurements of angular distributions (the angle is varied while keeping the beam energy constant) the uncertainty of the energy is totally correlated (i.e., systematic) for all measurements of one distribution, but this is not so for the uncertainty of the angle.
- For measurements of excitation functions (constant angle, beam energy varying) the uncertainty of the angle is totally correlated (i.e., systematic) for all measurements of one excitation function, but not the uncertainty of the energy. (If all energies are measured with the same instrument, a correlated component (e.g., the scale uncertainty of the instrument) exists for the energy uncertainty, too, but not the entire energy uncertainty is correlated.) Correlation could be entirely avoided by using instruments that are completely independent of each other – quite an unrealistic approach.

(2) Beam Properties

- mean particle energy (accelerator energy, energy loss),
- beam intensity (charge collection, charge integration),

The scale uncertainty of the charge integration is totally correlated for all measurements that use the same charge integrator; this will be true in most cases.

(3) Target Properties

- effective length,
- density of the target gas (purity, pressure, temperature).

There have been experiments where a different gas filling or an altogether different gas target was used for each measurement. Now, if the individual measurements of pressure and temperature, respectively, were done independently of each other, no systematic component would be present. However, if the same target with the same filling is used for all measurements, 100% correlation exists for the target uncertainties.

(4) Detector Properties

- beam direction (uncertainty in the zero point of the angle measurement),
- position (uncertainty of the angle measurement itself),
- detection probability (including solid angle subtended by the detector, detector bias stability, in- and out-scattering and attenuation of neutrons).

As discussed in *(1)*, it depends on the measurement circumstances whether a given angular uncertainty is correlated (systematic) or uncorrelated.

- When measuring an excitation function the *entire* uncertainty of the angle is correlated, i.e., systematic.
- For measurements of angular distributions a correlation exists predominantly for the uncertainty of the (mechanical) zero-point setting. If the zero point was set anew before each single measurement, this component would not entirely be systematic (to avoid correlation entirely independent determination methods would be needed).

To avoid systematic uncertainties of the detection probability one would need a different detector for each measurement – with their detection probability determined independently of each other.

(5) Counting Specific Properties

- "statistical" uncertainty,
- correction for lost counts,
- background subtraction.

The "statistical" uncertainty of counting is always uncorrelated; the uncertainty of the correction for count losses, though, will have a systematic component. This is also true for the background determination.

Summary

From the above discussion we have seen that the correlation among uncertainties depends on the type of experiment (excitation function, angular distribution). After the first measurement both possibilities are still open (it is still possible to keep the angle fixed and to vary the energy, or vice versa), therefore the following conclusions are necessarily true (see also the examples in Sect. 7.2.2):

1. No correlated uncertainty can exist for a single measurement.
2. Correlated and uncorrelated uncertainties are of the same nature, as an identical uncertainty can be correlated or uncorrelated, depending on the type of experiment.

It is important that uncertainty components that are independent of each other are added quadratically. This is also true for *correlated* uncertainty components, provided they are independent of each other, i.e., as long as there is *no correlation between the components.* If they depend on each other, their sum is calculated following the rules listed in Chap. 8. So one arrives at *combined uncorrelated* and *combined systematic (correlated) uncertainties* in the final data reduction of the experiment. Combining these two quantities to give a total uncertainty (as in Sect. 8.2.3) is not always the best action to take.

7.2.4 External Scale Uncertainties?

From a pattern of data values one cannot conclude how strongly the scale of the pattern fluctuates, i.e., how large the scale uncertainty is. Therefore, *external systematic uncertainties cannot be determined* directly. However, as we have shown in the cross-section measurement example above, the identical uncertainty can be viewed at as correlated or uncorrelated depending on the application. Therefore, an *uncorrelated* uncertainty induced from a pattern in one application can be taken over as a *correlated* uncertainty for another application, if identical.

Although it is not possible to gain information on the systematic uncertainty from the scatter of the data values an inductive approach to systematic uncertainties is often taken. In essence this implies that if an uncertainty does not behave like a random one it is systematic. Such a procedure of defining the systematic uncertainty inductively (based on external characteristics) and not on correlation is bound to fail. This absence of external systematic uncertainties could be the reason for the frequent miscomprehension (Sect. 7.3) of the nature of a systematic "error".

Problem

7.5. With a certain setup for counting nuclear radiation $N_a = 10{,}000$ counts are recorded during the measurement time $t_m = 10.000\,\mathrm{s}$. The *total* dead time t_D (Sect. 4.1.2) of this measurement was determined as 0.100 s (this was done by measuring the live time $t_a = 9.900\,\mathrm{s}$, using the same time base as for t_m). The uncertainty of the time measurement is basically given by the fractional uncertainty of the time base of $\pm 1 \times 10^{-4}$. The uncertainty of the dead time correction is negligible (Sect 4.1.3).

(a) Give the actual number of events N and the event rate $ER = N/t_m$.
(b) What are their (relative) uncertainties?

Note: A similar case but with different circumstances is discussed in Sect. 9.3.

7.3 Differentiation From "Systematic Errors"

The term *systematic error* is widely used ambiguously, be it by scientists in their work or by authors of books. Therefore, it is necessary to adhere to the term correlated or systematic *uncertainty* for better clarity. Let us sidetrack by leaving the subject of uncertainties for a discussion of *systematic deviations*. These are actually often meant when the term "systematic errors" is used.

7.3.1 Gross Mistakes

Gross mistakes are not uncertainties, but they are human errors when collecting data, e.g., in an experiment (insufficient understanding of the apparatus, taking readings incorrectly, mistakes when evaluating or interpreting the measured data). Omitting necessary corrections (Sect. 7.3.2) are also mistakes of that type. Gross mistakes should not happen, for they can be avoided when extra care is taken (double-checking, data redundancy, controls, scrutinizing, looking for ambiguous data). Unfortunately, this is not always done, and a lot of false data find their way into literature. If such a gross mistake has only been made once, i.e., for one point in a series, it may result in an outlier (Sect. 3.2.7).

7.3.2 Corrections

In Sect. 3.2.3 the need for corrections is amply exemplified. It should be quite obvious that also corrections have uncertainties (e.g., the dead time correction factor in Sect. 4.1.2). Such an uncertainty may be partially correlated. This correlated component in the uncertainty of a correction may be another reason for the frequent intermingling of the terms "systematic deviation" and "systematic uncertainty". In the following the consequences of not correcting are discussed.

Missing Corrections

In many cases *systematic errors* are interpreted as the systematic difference between *nature* (which is being questioned by the experimenter in his experiment) and the *model* (which is used to describe nature). If the model used is not good enough, but the measurement result is interpreted using this model, the final result (the interpretation) will be wrong because it is biased, i.e., it has a systematic deviation (not uncertainty). If we do not use the best model (the best theory) available for the description of a certain phenomenon this procedure is just wrong. It has nothing to do with an uncertainty.

Unknown influences on a measurement result can, obviously, not be taken into consideration and can therefore not be included in an uncertainty. On the other hand, *known* influences can be corrected because their size and sign are known. Not correcting does not cause a systematic uncertainty, but a systematic deviation, i.e., just a wrong result.

Examples

1. *Loaded Voltage Divider*. The voltage V_x at a voltage divider is to be measured. The voltmeter loads the voltage divider (drawing additional current), and the measured voltage V_m is always smaller than the voltage V_x to be measured, therefore V_m has to be corrected to give V_x. If this

correction is omitted and $V_x = V_m$ is used instead, the results deviate *systematically*. That is, they are simply *wrong* because the necessary correction has been omitted. The uncorrected value deviates in one direction (here it is too small), therefore the sign of the deviation is known. Thus we are dealing with a *systematic deviation* and not with a *systematic uncertainty*. Nevertheless, we often encounter the term "systematic error" for such circumstances (see also the example in Sect. 3.2.4).

2. *Buoyancy*. Naturally, the actual air pressure influences the correction for the buoyant force when weighing an object in air. This has to be considered in precise weight measurements. If the "model" used by the experimenter does not consider this dependence on the air pressure, which often is the case, the final result will be unnecessarily wrong since this effect is known and can be corrected. (In many cases, though, this effect is negligible.) However, unlike Example 1, the value of the correction changes in time due to the actual air pressure at the time of the measurement. Thus, using the nominal local air pressure corrects the buoyancy to a first order only. If the actual air pressure values at the time of the measurements have not been recorded, we might still be able to improve the correction by learning from a meteorological station nearby (after correcting for the difference in altitude) the mean air pressure value and its fluctuation over the measuring period. This provides not only the appropriate correction factor but also its uncertainty. As this uncertainty can be expected to be negligible when compared with the dominant one, its crude determination is insignificant.
3. *Temperature Dependence*. The voltage across a semiconductor diode is measured as $V_D = (0.653 \pm 0.004)\,\text{V}$ at a forward current of $I_D = (1.00 \pm 0.01)\,\text{mA}$. Now, the experimenter did not think about the voltage dependence on the ambient (air) temperature. Therefore he refrained from measuring the temperature at his working place. The dependence on temperature, though, is not negligible – it amounts to $-2\,\text{mV}/°\text{C}$. What can we do now? The experimenter really does not know the actual ambient temperature. But, he remembers that the room temperature was pleasant, if not a little warm during his experiment. He is sure that this temperature was less than 25°C and greater than 19°C, thus he assumes an ambient temperature of $(22 \pm 3)°\text{C}$ for this measurement. It is not possible for him to determine whether this value is too large or too small. (This type of uncertainty is naturally rather arbitrary; its value is more like a maximum uncertainty and in no case does it define a 1σ confidence interval.)
   Now there are two possible ways of including this systematic error (a missing temperature measurement) into the final result:
   - $V_D = (0.653 \pm 0.004)\,\text{V}$ for $(1.00 \pm 0.01)\,\text{mA}$ and $(22 \pm 3)°\text{C}$,
   - $V_D = (0.653 \pm 0.010)\,\text{V}$ for $(1.00 \pm 0.01)\,\text{mA}$ and 22°C.

In fact, we have no other choice than to treat this estimated uncertainty as a real one. Not considering the temperature dependence would be systematically wrong as it is known that it exists.

4. *Dead Time.* Electronic instruments for signal analysis have a so-called dead time (Sect. 4.1.2), i.e., the time they need for analyzing the signals. If the time intervals between all signals are greater than the dead time (periodical sequence of signals), no signals occur during the dead time and it is of no consequence. Radiation events originating from radioactive decay occur randomly, and short time intervals are exponentially favored (Sect. 4.1). Therefore the best estimates based on counting of such events are simply wrong (too small), if they are not corrected for dead time losses. On the other hand, correcting the measuring time for dead time when measuring periodic signals with a period that is longer than the individual dead time is just as wrong (Sect. 4.1.3) because no signals get lost!
5. *Gravitational Force of the Moon.* At CERN – one of the world's largest research laboratories, situated near Geneva – the influence of the gravitational force of the moon on the geometry of (circular) accelerators has been observed. The position of the moon can be deduced from the data. Here, in principle, we are dealing with the same situation: *Systematic deviations*, erroneously called systematic errors are caused if the data evaluation is based on a model that does *not* pay heed to geometric distortion as consequence of the mass attraction by the moon. Once again, this has nothing to do with an uncertainty and even less with a measurement uncertainty, but it indicates that the model (the theory) used was not suited for this problem, in fact, that it was too simple.

Faulty Corrections

This is a special case of systematic errors: corrections have been applied to the data, but this has been done systematically wrong, i.e., all data points deviate in the same direction. It is irrelevant whether this was done because of laziness or ignorance.

In particle physics there are abundant examples of such "laziness", not only in the years before digital computers were readily available (before about 1960). In those years the data conversion from the laboratory system to the center-of-mass system was done without taking relativistic effects into account. However, center-of-mass cross-section data involving protons with energies above about 10 MeV differ markedly when evaluated correctly (relativistically) from those evaluated the easy way (nonrelativistically).

Even if the corrections of all data values of a data set are wrong the same way, i.e., if they are systematically wrong, this does not establish a systematic uncertainty. These deviations could have been avoided by more careful corrections. Thus we encounter deviations of badly corrected values from "correctly" corrected values that are by no means uncertainties.

Disregarding Corrections

If there is a difference in the quality of an empirical data value and its description by theory then the data value must be corrected for this difference (or the theory changed). There is *just one* (very good) excuse for omitting a correction. If a correction affects *none* of the significant figures of a best estimate it is obviously superfluous.

Problems

7.6. A gamma ray spectrum (from a radioactive source) has been recorded (in the live time mode, see Sect. 4.1.3). The background is assumed to be given by a straight line, determined by the two border points (X_1 and X_2).

(a) Determine the number of counts N originating from the source and
(b) its uncertainty ΔN.
The integral N_t of all events between the border points (channel numbers $X_1 = 662$ and $X_2 = 765$) is $N_t = 20{,}258{,}830$ and the individual counts A_1 and A_2 at the border points are 66,403 and 44,107.
Note: Disregard the instability of the pulse height (= x-axis), therefore $\Delta X_\mathrm{i} = 0$.
(c) Which uncertainty component is dominant?

7.7. Do parts (a), (b) and (c) as in Problem 7.6, but with subtraction of a physically relevant background N_{bg} (the Compton edge) that was simulated giving the value $N_{bg} = 1{,}675{,}843$ in the pulse height range (662–765).

(d) Which of the two background determinations (7.6 or 7.7) is of higher precision?
(e) Which of the two background determinations is of higher accuracy (by assumption)?

Note:

- For this case it is difficult to estimate the uncertainty of the background correction. It is slightly subjective and could be made more objective, e.g., by measuring a known relation. A 20% uncertainty in N_{bg} should be viewed as conservative (i.e., at any rate large enough) if done by an experienced experimenter.
- In Problem 7.6 the simplified assumption of a linear background results in a best estimate that is (systematically) too small, i.e., just wrong.

7.4 Correlation in Cases of Linear Regression

The situation for external uncertainties is quite different from that for internal uncertainties. External uncertainties are determined inductively by comparing data values with their functional relation. It is not possible to determine external uncertainties without a functional relation.

Table 7.1. List of $n = 3$ data values with weights

	x_i	y_i	x_i^2	y_i^2	$x_i \cdot y_i$	w_i
	−1.	0.8816	1.	0.7773	−0.8816	50.3
	−2.	2.1191	4.	4.4906	−4.2382	100.0
	−3.	2.8817	9.	8.3045	−8.6451	50.3
Sum	−6.	5.8824	14.	13.5724	−13.7649	200.6
Mean value	−2.	1.9608				

Consequently, the best estimate of data values with external uncertainties will be given by way of parameters of a functional relation. Only for a one-parameter presentation (e.g., the mean value) could the best estimate be interpreted as an "improvement" of the data values. As a consequence, no correlation can exist between external uncertainties; correlation can only exist between the parameters, or between the data and the function by which they are described. This fact is discussed in more detail in Sect. 7.4.2.

7.4.1 Weighted Linear Regression (Example)

The weighted linear regression was introduced in Sect. 6.3.2. It is an attempt to find the "best" straight line, i.e., the best parameters a_0 and a_1 of the equation of a straight line

$$y = a_1 \cdot x + a_0 \tag{7.11}$$

that best presents n data points of the values y_i and their internal uncertainties Δy_i. It is required that the x_i values have "no" uncertainty (or, better, a negligible uncertainty). As in Sect. 6.3.1, each point is assigned a weight factor w_i:

$$w_i = \frac{1}{(\Delta y_i)^2} \, . \tag{7.12}$$

Table 7.1 presents a simple example. Three points $(x_i,\ y_i)$ are given with their weights w_i. For them the best estimate (parameters a_0 and a_1) should be determined by least-squares fitting a regression line. For the calculation of the parameters a_0 and a_1 and their standard deviations $\sigma_{a0} = \Delta a_0$ and $\sigma_{a1} = \Delta a_1$, it is practical to introduce the following abbreviations (in each case the sum is taken from 1 to n):

$$A = \sum w_i x_i = -401.2 \, , \tag{7.13}$$

$$B = \sum w_i = 200.6 \, , \tag{7.14}$$

$$C = \sum w_i y_i = 401.2 \, , \tag{7.15}$$

$$D = \sum w_i x_i^2 = 903.0 \, , \tag{7.16}$$

$$E = \sum w_i x_i y_i = -903.0 \, , \tag{7.17}$$

$$G = D \cdot B - A^2 = 20180.4 \,. \tag{7.18}$$

The equations from Sect. 6.3.2 can now be written as

$$a_0 = (D \cdot C - E \cdot A)/G \,, \tag{7.19}$$
$$a_1 = (E \cdot B - C \cdot A)/G \,, \tag{7.20}$$
$$\sigma_{a0}^2 = D/G \,, \tag{7.21}$$
$$\sigma_{a1}^2 = B/G \,, \tag{7.22}$$
$$s_{01} = -A/G \,. \tag{7.23}$$

Thus for the parameters of the line through the three data points we get $a_0 = 0.000 \pm 0.212$ and $a_1 = -1.000 \pm 0.100$, making the equation of the line $y = -x$.

In addition, we get the variances $\sigma_{a0}^2 = 0.0447$ and $\sigma_{a1}^2 = 0.0099$, and also the covariance $s_{01} = 0.0199$. The latter shows that the correlation between a_0 and a_1 is obviously small.

7.4.2 Linear Regression Without Weighting (Example)

For the three data values in Table 7.2 no internal uncertainties are given, therefore all points have the same weight w, the inverse of the variance σ_y^2 of the values y_i:

$$w = \frac{1}{\sigma_y^2} \,. \tag{7.24}$$

As all data values have the same weight factor w the abbreviations introduced in Sect. 7.4.1 can be simplified as follows (again all sums to be taken from 1 to n):

$$A = \sum x_i \,, \tag{7.25}$$
$$B = \sum 1 = n \,, \tag{7.26}$$
$$C = \sum y_i \,, \tag{7.27}$$

Table 7.2. List of $n = 3$ data values without weights

	x_i	y_i	x_i^2	y_i^2	$x_i \cdot y_i$	q_y
	−1.	0.92	1.	0.8464	−0.92	0.0064
	−2.	2.16	4.	4.6656	−4.32	0.0256
	−3.	2.92	9.	8.5264	−8.76	0.0064
Sum	−6.	6.00	14.	14.0384	−14.00	0.0384
Mean value	−2.	2.00				
Variance	1.	1.0192				0.0384
Std. dev.	1.	1.0096				0.196

$$D = \sum x_i^2 \,, \tag{7.28}$$

$$E = \sum x_i y_i \,, \tag{7.29}$$

$$F = \sum y_i^2 \,, \tag{7.30}$$

$$G = D \cdot B - A^2 \,, \tag{7.31}$$

to accommodate the equations of Sect. 4.2.3 as follows:

$$a_0 = (D \cdot C - E \cdot A)/G \,, \tag{7.32}$$

$$a_1 = (E \cdot B - C \cdot A)/G \,, \tag{7.33}$$

$$\sigma_{a0}^2 = D/(w \cdot G) \,, \tag{7.34}$$

$$\sigma_{a1}^2 = B/(w \cdot G) \,, \tag{7.35}$$

$$s_{01} = -A/(w \cdot G) \,. \tag{7.36}$$

The value for w results from the sum of the squares of the deviations q_y and amounts to $w = 1/0.0384$. The loss of two degrees of freedom (Bessel correction) due to the calculation of the two parameters a_0 and a_1 has been taken into account.

The value of w is not used for the calculation of a_0 and a_1. Consequently one gets the identical regression line independent of the uncertainty value, as long as all data values have the same one, be it external or internal.

Thus the parameters of the line through the three data points equal $a_0 = 0.000 \pm 0.299$ and $a_1 = -1.000 \pm 0.139$, and the equation of the line is $y = -x$.

In addition, we get the variances $\sigma_{a0}^2 = 0.0896$ and $\sigma_{a1}^2 = 0.0192$, and also the covariance of these two coefficients $s_{01} = 0.0384$. As discussed in Sect. 4.2.3, the parameter a_0 is only relevant together with a_1. The covariance s_{01} is a measure of the mutual dependence of these two parameters.

After having studied the correlation between the coefficients of the equation we now want to investigate the linear relation between the data values (via the product–moment correlation after Bravais–Pearson, see Sect. 7.1.1). The variance amounts to

$$\sigma_x^2 = (D - A^2/n)/(n-1) = 1 \,, \tag{7.37}$$

and

$$\sigma_y^2 = (F - C^2/n)/(n-1) = 1.0192 \,, \tag{7.38}$$

and the covariance is obtained as

$$s_{xy} = (E - A \cdot C/n)/(n-1) = -1 \,. \tag{7.39}$$

With

$$r_{xy} = s_{xy}/(\sigma_x \cdot \sigma_y) = -0.9905 \,, \tag{7.40}$$

we get the linear correlation coefficient. This coefficient states how well the coordinates (x_i, y_i) agree with their presentation as a line, i.e., how far the value of y_i is fixed by the knowledge of x_i. If Pearson's correlation coefficient r_{xy} equals 0, the coordinates have nothing in common with a line. If $r_{xy} = 1$ is true, all points are on a line with a positive slope, i.e., the x_i and y_i grow equidirectionally; if $r_{xy} = -1$ is true, the x_i and y_i are negatively correlated. Here we obviously encounter a negative correlation, because y decreases with increasing x values.

Significance of a Correlation Coefficient

The above value of $r_{xy} = s_{xy}/(\sigma_x \cdot \sigma_y) = -0.9905$ seems to signify a highly significant linear correlation between the three points. However, not surprisingly the significance depends strongly on the number of points involved. The above three points could just as well be part of an uncorrelated set of data points; their correlation coefficient close to 1 could be chance. From appropriate tables it can be learned that there is an 8% chance that these three data points are uncorrelated. If a correlation coefficient of the same value were obtained from, e.g., five data points, the probability of a chance correlation would only be 0.4%. A correlation coefficient of, e.g., $r_{xy} = 0.7$ becomes significant (i.e., 5% chance correlation) only if it is based on at least eight data points.

Taking two data points of any bivariate data set will automatically result in $r_{xy} \equiv 1$, as can easily be found out by choosing $n = 2$ in the above equations. This is not surprising because any two points lie on a straight line. Consequently the significance of such a correlation is zero even if the linear correlation coefficient equals one.

Problems

7.8. Verify the numerical answers given in the examples of Sects. 7.4.1 and 7.4.2.

7.9. Does $r_{xy} = -0.6$ reflect a stronger linear relationship than $r_{xy} = 0.5$?

7.10. Women living in a rooming house were checked in the course of a mass health screening: among other factors, their blood pressure (under equivalent circumstances) was recorded. From these measurements 15 were (randomly) chosen; the data pairs – systolic blood pressure value and age – are listed in Table 7.3.

(a) Is the assumption legitimate that there is a (linear) correlation between these two values?
(b) If there is a linear correlation determine the two parameters of the line.

Table 7.3. Record of systolic blood pressure value and age of 15 women

Pressure	157	139	142	129	133	136	119	112	114	118	120	134	145	141	145
Age	62	51	48	46	43	40	34	29	31	37	40	45	48	50	58

7.5 Data Consistency Among Data Sets

When comparing data sets or some (evaluated) best estimates, attention has to be paid to correlation between the individual data sets and their uncertainties, respectively. Special care should be taken to quote the systematic and the uncorrelated components of the total uncertainty separately (Sect. 8.2.3). Furthermore, it makes sense first to combine those data sets that were acquired by the same method before the final evaluation is started. Data that were gained the same way could be subject to analogous mistakes (like omitted or incomplete corrections); combining data obtained by different methods would veil such (possible) problems.

Example. ***Neutron Counter Telescope***

> Measurements of neutron sources using neutron counter telescopes have often resulted in data that were too small by 4–10% when compared with data measured with other methods. For those cases where the neutron source is cooled with water spray, there is an obvious explanation: the subtracted background was (systematically) too large because protons were knocked out from the cooling water spray. Because of the radiator these protons would not be counted in the foreground measurement, but only in the background measurement.

Evaluation means extracting best estimates from data sets of different origins. To do so, we need to find out

- which data sets are consistent with each other,
- which data sets depend on each other, i.e., are correlated. Not recognizing dependences will put too much weight on such correlated data sets.

Determining consistency *between* data sets is more complicated than determining consistency of data values *inside* a data set (Sect. 6.4), since we usually do not have as many data sets available as data values inside a data set. If just two (inconsistent) data sets exist, there is, in general, no way of telling which of these two should be chosen as best estimate (if either of the two is suited for this at all). In general the procedure is quite similar to that discussed in Sect. 3.2.7. As a first step one would proceed following the checklist there, but finally one would be forced to compromise and to determine the results using both sets, even knowing that this gives a result of less accuracy than the result of the correct data set (that might be there). So, data evaluation sometimes is a matter of luck and of experience. *What remains important is proceeding correctly and documenting the steps taken.*

The correct procedure might consist in, for instance, rejecting a third data set if two obviously independent and consistent data sets exist, and the third data set is obviously not consistent with the others. It has, nevertheless, happened that this correct procedure has given the wrong best estimate: The rejected data set was "right" and the two other data sets were wrong – this fact could only be shown after additional data had become available.

With the massive use of computers in this field the relation among data sets is described by so-called *correlation matrices* (see also Sect. 8.3.3). Below we present two, hopefully inspiring examples of actual evaluations.

7.5.1 Contradictory Data Sets

Example. *Consistency of Data Sets*

In the year 1973 a thorough evaluation of the cross sections of the reactions ^{3}H(p, n)^{3}He, ^{2}H(d, n)^{3}He, and ^{3}H(d, n)^{4}He up to projectile energies of 10 MeV was published (LI73, Liskien H, and Paulsen A (1973) Nuclear Data Tables A11: 569). This evaluation was based on data that had been published up until December 1972. In the following the same data identifiers are used as in this evaluation.

Let us have a closer look at a small portion of this evaluation, namely the neutron production at 0° by the reaction ^{3}H(d, n)^{4}He for deuteron energies between 7 and 10 MeV. In this energy range six data sets were available; one of these (BR51, 9.8 MeV) was rejected by the evaluators. The data of the remaining five sets show no unreasonable scattering, so a solution was found, obviously favoring the most recent data set (Si68) that was not in disagreement with the other four data sets. Actually, the data situation was not satisfactory to the evaluators as indicated by their comment: "especially at energies above 5 MeV, further investigations would be worthwhile".

After including many of my own data (DR78) my evaluation of these reactions for projectile energies up to 16 MeV was published in the year 1987 (DR87) – an evaluation that is still valid today with a few adjustments later on. What has changed for neutron production cross sections at 0° by the reaction ^{3}H(d, n)^{4}He for deuteron energies between 7 and 10 MeV by this new evaluation, and what are the reasons for the discrepancies?

The largest change concerns the end point of the previous evaluation at 10 MeV; its value with an uncertainty of ±4% was shown to be low by 10%. Rigorous sighting of the data originally available, but with new insights, gives the following results:

- The data set Si68, which is obviously the foundation of the evaluation, contained only preliminary data that had never been published because their values were (mysteriously) low by about 6–10%, not unlike other measurements using neutron counter telescopes, as discussed in the example given at the beginning of this section.

- The data sets GO61A and BR64 that were used in the evaluation are (both) relative data sets; they do not contain absolute values. Therefore they are not in disagreement with a best estimate at 10 MeV that is higher by 10%.
- The data value at 7.3 MeV of BA57C is, on the one hand, susceptible to mistakes because it is the high-energy end point of the data set. On the other hand, its value is not really crucial in determining the 10-MeV value.
- The data ST60 between 6.1 and 14.2 MeV that were measured with nuclear photographic plates support the new evaluation after applying appropriate corrections. They are based on the cross-section standard ^{1}H(n, n)^{1}H and were given in the center-of-mass system. The nonrelativistic transformation of the measured data from the laboratory system to the center-of-mass system resulted in faulty values of the angles as well as of the cross sections. After appropriate corrections for a better standard and for the relativistic effects these data were consistent with the new data (DR78).
- Finally, it was shown that the data value at 9.8 MeV (BR51) that had previously been rejected is also consistent with the new best estimate. Figure 7.1 gives an impression of the discrepancy between the two evaluations.

Even though no inconsistency was detected in the evaluation from 1973 for the five data sets used (just an unsatisfactory situation, see above), the best estimate (at 10 MeV) determined later was higher by 2.5 standard deviations. From this we can tell that *consistency alone* cannot guarantee a reliable best estimate: consistency is *in no way sufficient* for the success of an evaluation.

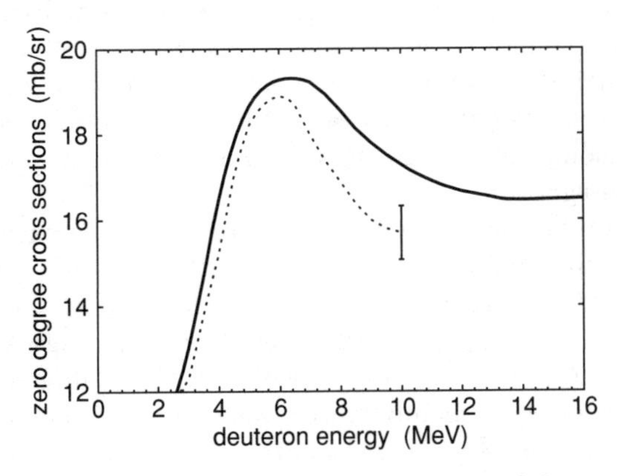

Fig. 7.1. Evaluation of the center-of-mass cross sections of the neutron production at 0° by the reaction ^{3}H(d, n)^{4}He. *Solid line* from DR87, *dashed line* from LI73

Neither is there a foolproof technique for handling discrepant data. Consequently, a personal touch of the evaluator will always remain.

7.5.2 Dependent (Correlated) Data Sets

When a best estimate is derived from a number of data sets, it is vital to determine which of these sets are in fact independent, so that the weighting (when combining these sets) can be correctly done. Otherwise too much weight in the evaluation is laid on the dependent data sets that, in reality, might just be one (independent) data set.

Example. ***Dependent Data Sets***

Once again let us choose data of the reaction ^{3}H(d, n)^{4}He, this time its total cross sections. In the year 1991 the LLNL evaluation of this reaction was presented at an international conference. Table 7.4 compares the best estimates of the 1987 evaluation by Drosg – since 1987 available for the general public in the form of a computer code – and the LLNL evaluation presented in 1991. The latter claims to be an independent evaluation that uses the extrapolation method in cases (as here for 19.00 MeV) where no data were available.

The first thing noted is that the mean value of all the ratios is 1.0002 (see Column 4), i.e., that the difference in scale equals $(0.02 \pm 1.80)\%$, averaged over 21 data values in the quite substantial energy range of 13 MeV. This uncertainty is the probable (mean) uncertainty based on the specification of the accuracy of the data ($\pm 1.5\%$ for σ_D, $\pm 1.0\%$ *for* σ_L) in both evaluations. (The contribution of the external uncertainties is negligible.) The probability that such a scale difference of 0.0111 standard deviations will occur between two independent data sets is 0.89% (Table 5.3); this means that the agreement of the scales is *highly significant*, i.e., the scales are strongly correlated.

Investigating the individual differences among the 21 data values that are quite evenly distributed in the deuteron energy range of 13 MeV, we find that the largest difference is only 0.29%, even though the total cross sections were derived from incomplete angular distributions – the data were calculated by extrapolating integration.

One of the reasons for the small, but detectable difference between these two data sets is the rounding effect for the DROSG87 data (which only contain significant digits, other than the LLNL data). Is it then legitimate to assume that the difference is mainly caused by rounding? We need to determine whether the difference in the ratios of the evaluated cross sections agrees with a model that mainly traces this difference back to the rounding uncertainty. For making such a decision the chi-squared test (Sect. 8.3.3) can be used in which the squares of the deviations (as a measure of the external uncertainties) are brought into relation with the squares of the internal uncertainties. Following the lines of our assumption, the rounding uncertainty is used as the internal uncertainty. Since the scale has not been adjusted, no

Table 7.4. Comparing evaluated data of the total cross section of the reaction ^{3}H(d, n)^{4}He. Data of the computer code DROSG87 and the LLNL evaluation are compared. The following data are also given: the ratio σ_D/σ_L minus 1 of the individual data, and the square of the deviations q, the square of the rounding uncertainty Δy^2, and their ratio $q/\Delta y^2$. Note that here the symbol σ stands for the cross section

E_d (MeV)	σ_D(DR87) (barn)	σ_L(LLNL91) (barn)	$\sigma_D/\sigma_L - 1$	q (10^{-4})	Δy^2 (10^{-4})	$q/\Delta y^2$
6.00	0.0765	0.076534	−0.0004	0.0016	0.0043	0.372
6.20	0.0748	0.074944	−0.0019	0.0361	0.0045	8.022
6.50	0.0724	0.072412	−0.0002	0.0004	0.0048	0.083
7.00	0.0683	0.068331	−0.0005	0.0025	0.0054	0.463
7.50	0.0649	0.064856	0.0007	0.0049	0.0059	0.831
7.90	0.0624	0.062311	0.0014	0.0196	0.0064	3.063
9.10	0.0556	0.055555	0.0008	0.0064	0.0081	0.790
10.00	0.0515	0.051492	0.0002	0.0004	0.0094	0.043
10.70	0.0491	0.048980	0.0025	0.0625	0.0104	6.010
11.00	0.0481	0.048023	0.0016	0.0256	0.0108	2.370
11.40	0.0468	0.046825	−0.0005	0.0025	0.0114	0.219
12.00	0.0451	0.045119	−0.0004	0.0016	0.0123	0.130
12.30	0.0442	0.044310	−0.0025	0.0625	0.0128	4.883
13.00	0.0425	0.042533	−0.0008	0.0064	0.0138	0.464
13.36	0.0417	0.041694	0.0001	0.0001	0.0144	0.007
14.00	0.0404	0.040305	0.0024	0.0576	0.0153	3.765
14.20	0.0399	0.039900	−0.0000	0.0000	0.0157	0.000
15.00	0.0385	0.038387	0.0029	0.0841	0.0169	4.976
16.00	0.0367	0.036720	−0.0005	0.0025	0.0186	0.134
16.50	0.0360	0.035968	0.0009	0.0081	0.0193	0.420
19.00	0.0328	0.032823	−0.0007	0.0049	0.0232	0.211
Sum			0.0051	0.3903	0.2437	37.256
Mean value			0.0002	0.0186	0.0116	1.774

degree of freedom was lost, so we need to divide by the number of points $n = 21$.

The mean value of the last column gives the value of chi-squared, namely 1.8. This value is close enough to the expected value (i.e., 1.) to support the assumption that the rounding effect is mainly responsible for the very small difference.

Evaluations of the same quantities will generally be correlated in some way because they are based on (nearly) the same reservoir of data. If, as in above example, a later evaluation that, in addition, uses fewer data values agrees in a highly significant way with a previous one it looks as if this portion of the two evaluations is identical. In such a case it would be natural to find a correlation of basically 100%.

Problems

7.11. What is the probability that a normally distributed value will deviate by more than 2.5 standard deviations from the true value?

7.12. Verify that an agreement between two data sets is highly significant if the mean of all ratios is (1.0002 ± 0.0180). What is the probability that a data value will deviate less than 0.0111 standard deviations from the true value?

7.6 Target Shooting as a Model for Uncertainties

Target shooting can be a model used to explain some of the technical terms we have already encountered. I apologize to marksmen for this very basic approach of a layman – this section does not deal with shooting; it deals with uncertainties.

After production of a gun and its first use (this step corresponds to artificial aging or thermal cycling in electronic instruments), the sight will need to be adjusted. After these adjustments the *mean deviation* of the hits on the target to the point aimed at (e.g., the bull's-eye) will be smaller than a certain angle. This deviation corresponds to the *systematic* (or *correlated*) *scale uncertainty*.

With the weapon's position fixed, for instance, in a vise, in such a way that it is aimed at a target, hits will scatter around a mean point corresponding to the inevitable uncorrelated uncertainty intrinsic to the instrument. These variations are mainly due to the tolerance in the production process of the cartridges and the individual (random) radial positioning of the cartridges in the barrel. (This corresponds to the *uncorrelated* portion of the *scale uncertainty.*)

Another cause for deviations is the wind. If the direction and speed of the wind during shooting can be determined, a *correction factor* can be introduced that will result in an offset when aiming. Sometimes the direction of the wind changes too quickly for corrections to be effective, so the deviations of the hit positions from the desired aim become greater. This effect can be described by a further *uncorrelated uncertainty.*

In reality the weapon is hand-held, which has the following consequences:

- Even small hand movements result in movements of the weapon. These movements cannot be predicted; they are random. These movements, too, cause additional scatter corresponding to an additional *uncorrelated uncertainty*.
- Differences in the way the weapon is held, for instance, if it is slightly tilted when compared to the orientation when fixed in a vise, result in a systematic deviation from the desired aim. This effect can be corrected by aiming at an appropriate point next to the bull's eye, i.e., a *correction factor* must be applied.

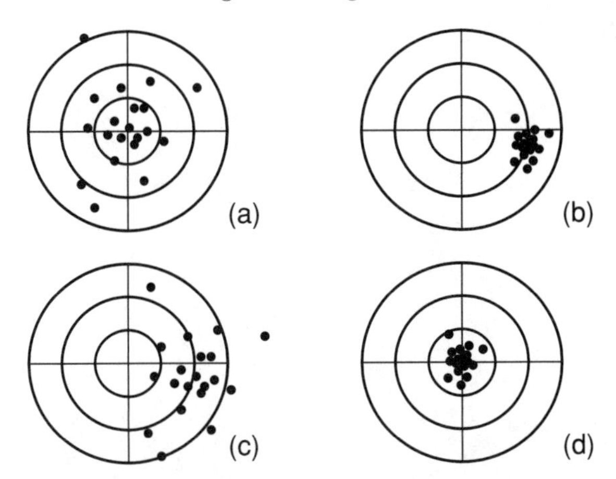

Fig. 7.2. Accuracy vs. precision: (**a**) low precision, high accuracy; (**b**) high precision, low accuracy; (**c**) low precision, low accuracy; (**d**) high precision, high accuracy

- Hits that are far off are equivalent to *outliers*. In most cases these are caused by a mistake of the person shooting (or in the case of an experiment, the experimenter). Other possible causes include faulty cartridges, etc.

If many different people use the same weapon to shoot at their individual targets, the shooting patterns can be compared and analyzed to separate the individual shooting pattern from that intrinsic to the gun (the intrinsic *uncorrelated uncertainty*).

On the other hand, if a person with constant (good) shooting performance uses different guns under otherwise identical shooting conditions, we can analyze these shooting patterns to get information about the aiming properties of the weapon (*correlated uncertainty*).

The terms *precision* and *accuracy* can also be visualized in this context: If the hit area on the target is small, high precision has occurred; if the center of the hit area is very close to the bull's-eye, high accuracy is the result. Quite similar to science (Sect. 10.1), a balance between precision and accuracy is quite important in target shooting, too. If one dominates the result will not be satisfactory in either case. Furthermore, a single shot resembles a single data value; the distinction between accuracy and precision gets lost.

However, target shooting does not entirely reflect the situation encountered with uncertainties. The center of the target is not a *true value* but rather a nominal value like that in the commercial production of mechanical pieces. Figure 7.2 illustrates the two terms *accuracy* and *precision*.

8

Dealing With Internal Uncertainties

If the quantity F depends on a number of variables $x, y, \ldots$ with uncertainties $\Delta x, \Delta y, \ldots$, the influence of the individual uncertainty components on the total uncertainty ΔF depends on the type of mathematical function used to describe F. The formalism that shows how the uncertainties $\Delta x, \Delta y, \ldots$ are propagated and how the total uncertainty ΔF of the final result F is gotten is described by the *general law of error propagation.* It is sufficient to restrict ourselves to two variables (x and y) as this is a recursive law, i.e., by repeated application we get the corresponding result for any number of variables.

Let us first consider the situation with *external* uncertainties. The variance of F as presented by the *general law of error propagation* is given by

$$\sigma_F{}^2 \approx (\partial F/\partial x)^2 \cdot \sigma_x{}^2 + (\partial F/\partial y)^2 \cdot \sigma_y{}^2 + 2 \cdot (\partial F/\partial x) \cdot (\partial F/\partial y) \cdot s_{xy} \,. \quad (8.1)$$

This variance contains the variance of the two components and the partial derivatives of the dependence of F on x and y. Contrary to the quadratic addition presented in Sect. 3.4, a mixed term is present, containing the covariance s_{xy} (Sect. 7.4.2) in addition to the two dependences. This covariance is a measure of the relation between the two characteristics x and y (Sect. 7.4.2).

After substituting the covariance s_{xy} by the linear correlation coefficient r_{xy} (Sect. 7.4.2) one obtains

$$\sigma_F{}^2 \approx (\partial F/\partial x)^2 \cdot \sigma_x{}^2 + (\partial F/\partial y)^2 \cdot \sigma_y{}^2 + 2 \cdot (\partial F/\partial x) \cdot \sigma_x \cdot (\partial F/\partial y) \cdot \sigma_y \cdot r_{xy} \,. \quad (8.2)$$

In the limiting case of $|r_{xy}| = 1$, i.e., in the case of a complete correlation, the relation can be simplified to

$$|\sigma_F| \approx |(\partial F/\partial x) \cdot \sigma_x \pm (\partial F/\partial y) \cdot \sigma_y| \,, \quad (8.3)$$

with the sign of the second term equal to that of r_{xy}. This general law of error propagation (for *external* uncertainties) needs the linear correlation coefficient r_{xy} that is determined from the pattern of the data values.

In the case of *internal* uncertainties the standard deviations must be replaced by the values of the internal uncertainties, and instead of the correlation

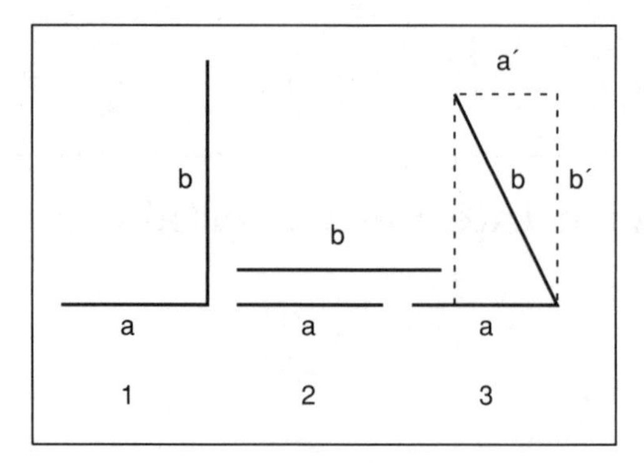

Fig. 8.1. Geometric visualization of the correlation between two uncertainties a and b. (**1**) No correlation, a and b do not have a common directional component. (**2**) Total correlation, a and b are equidirectional. (**3**) Partial correlation. The component a' of b that has the same direction as a can by determined by treating b as a vector and splitting it into its orthogonal components a' and b'

coefficient r_{xy} we introduce the factor k, the *degree of correlation* between the two uncertainty components Δx and Δy. Consequently the *general law of error propagation for internal uncertainties* becomes

$$(\Delta F)^2 = (\partial F/\partial x)^2 \cdot (\Delta x)^2 + (\partial F/\partial y)^2 \cdot (\Delta y)^2 + 2 \cdot (\partial F/\partial x) \cdot \Delta x \cdot (\partial F/\partial y) \cdot \Delta y \cdot k \,. \tag{8.4}$$

The correlation between Δx and Δy is mutual, i.e., Δx and Δy are equivalent (and without orientation). Consequently the correlation has no direction, so that k is always positive. An uncertainty does not have a sign; just its absolute value matters. The sign of the mixed term in the above equation is determined by that of the partial differentiation coefficients. This is why this degree of correlation – just like the degree of acoustic correlation – can be written as $k = \cos\phi$, where $0° \leq \phi \leq 90°$; this is consistent with the geometric interpretation given in Fig. 8.1.

As is the case with the correlation coefficient r_{xy} (Sect. 7.1.1), the square of the degree of correlation k^2 is introduced because it is more elucidative. If the correlated and the uncorrelated components of an uncertainty are of the same size, $k = 1/\sqrt{2}$, but $k^2 = 0.5$ (i.e., both contribute 50% each), as demonstrated in the following example.

Example. ***Partial Correlation***

Two uncertainties Δx and Δy be partially correlated, such that this partial correlation of Δy can be split into *two equally sized components* Δy_u (uncorrelated) and Δy_c (totally correlated). As the equation for the general error propagation is symmetric in x and y, and the correlation is mutual (i.e., relative to each other) it is not important whether Δx or Δy is split. From

$$(\Delta y)^2 = (\Delta y_u)^2 + (\Delta y_c)^2 = 2 \cdot (\Delta y_c)^2 \tag{8.5}$$

one obtains

$$(\Delta y_c)^2 = 0.5 \cdot (\Delta y)^2 , \tag{8.6}$$

so that the total *correlated* uncertainty component is given as

$$\begin{aligned}(\Delta F_c)^2 &= (\partial F/\partial x)^2 \cdot (\Delta x)^2 + (\partial F/\partial y)^2 \cdot (\Delta y_c)^2 \\ &\quad +2 \cdot (\partial F/\partial x) \cdot \Delta x \cdot (\partial F/\partial y) \cdot \Delta y_c \cdot k .\end{aligned} \tag{8.7}$$

Because of $k = 1$ this results in a total *correlated* uncertainty component of

$$\begin{aligned}(\Delta F_c)^2 &= (\partial F/\partial x)^2 \cdot (\Delta x)^2 + 0.5 \cdot (\partial F/\partial y)^2 \cdot (\Delta y)^2 \\ &\quad \pm\sqrt{2} \cdot (\partial F/\partial x) \cdot \Delta x \cdot (\partial F/\partial y) \cdot \Delta y .\end{aligned} \tag{8.8}$$

The variance of the total *uncorrelated* uncertainty component is

$$(\Delta F_u)^2 = (\partial F/\partial y)^2 \cdot (\Delta y_u)^2 = 0.5 \cdot (\partial F/\partial y)^2 \cdot (\Delta y)^2 . \tag{8.9}$$

The variance of the *total uncertainty* is obtained by addition of both components as

$$\begin{aligned}(\Delta F)^2 &= (\Delta F_c)^2 + (\Delta F_u)^2 \\ &= (\partial F/\partial x)^2 \cdot (\Delta x)^2 + (\partial F/\partial y)^2 \cdot (\Delta y)^2 \\ &\quad \pm\sqrt{2} \cdot (\partial F/\partial x) \cdot \Delta x \cdot (\partial F/\partial y) \cdot \Delta y .\end{aligned} \tag{8.10}$$

When comparing this result with the *general law of error propagation*, as shown in (8.4), it is found that here $k = (\pm)1/\sqrt{2}$. With k being positive this corresponds to $\phi = 45°$ or $k^2 = 0.5$, as is expected for two components of the same size ($\Delta y_u = \Delta y_c$).

Whereas the correlation coefficient r_{xy} takes care of the correlation between the data values in the case of (inductively) determined external uncertainties, the degree of correlation k is solely necessary to satisfy the general law of error propagation for *internal* uncertainties. Internal uncertainties are deduced from the internal properties of the uncertainty components. Therefore, it is feasible to select these components in a way that either the property "uncorrelated" or "fully correlated" can be assigned to each component in a straightforward way. For this reason only $k = 0$ and $k = 1$ need to be considered with internal uncertainties (as done, e.g., in Sect. 7.2).

In the case of fully *uncorrelated* uncertainties (where no dependence exists between Δx and Δy), $k = 0$, and the mixed term of the general law of error propagation drops out. We get the generally known equation for addition in quadrature, a special case of (3.13):

$$(\Delta F)^2 = (\partial F/\partial x)^2 \cdot (\Delta x)^2 + (\partial F/\partial y)^2 \cdot (\Delta y)^2 . \tag{8.11}$$

Thus *components $\Delta x, \Delta y, \ldots$ that are uncorrelated must be added quadratically.* Such components that are at most one third as large as the largest (dominant) component can be disregarded (Sect. 3.4.1). This is a great help if the quantity F is described by a complicated function that, in general, results in an even more complicated function for ΔF. Sometimes it is helpful to remember that the quadratic addition is also called geometric addition (Sect. 3.4).

Totally correlated uncertainties Δx and Δy are fully dependent on each other, and with $k = 1$ we get

$$(\Delta F)^2 = (\partial F/\partial x)^2 \cdot (\Delta x)^2 + (\partial F/\partial y)^2 \cdot (\Delta y)^2 + 2 \cdot (\partial F/\partial x) \cdot (\partial F/\partial y) \cdot \Delta x \cdot \Delta y \,. \tag{8.12}$$

Compared to the uncorrelated case there is an additional mixed term. However, this does not necessarily mean that the total uncertainty is larger for correlated components than for uncorrelated ones of the same size.

The above equation can be simplified to

$$(\Delta F)^2 = ((\partial F/\partial x) \cdot \Delta x + (\partial F/\partial y) \cdot \Delta y)^2 \,, \tag{8.13}$$

yielding

$$\Delta F = \pm[(\partial F/\partial x) \cdot \Delta x + (\partial F/\partial y) \cdot \Delta y] \,. \tag{8.14}$$

Thus *components $\Delta x, \Delta y, \ldots$ that are fully correlated must be added linearly.*

The (linear) addition of correlated components gives a larger result only when both partial derivatives have the same sign. Otherwise it results in a subtraction that can even cause the cancellation of correlated uncertainty components (see, e.g., Sect. 8.1.4).

8.1 Calculations With Both Types of Uncertainties

Values for the degree of correlation k between 0 and 1 are not used usually; uncertainties with such partial correlation can easily be split into correlated ($k = 1$) and uncorrelated ($k = 0$) components. This is the reverse of the procedure discussed in Sect. 8.2.3 that allows us to calculate a total uncertainty with the help of the uncorrelated and correlated uncertainties. The correlated and the uncorrelated components can be extracted by splitting the uncertainty, just as a vector can be split into its components (see also Fig. 8.1). Figure 8.2 gives the five solutions for the sum of the uncertainties of the three cases depicted in Fig. 8.1. In Fig. 8.2 the length of the line is a measure of the size of the combined uncertainty.

Whereas only one solution exists for the uncorrelated case (quadratic addition), two solutions exist for the correlated case (addition or subtraction), and also for partial correlation (addition or subtraction of the correlated components followed by quadratic addition).

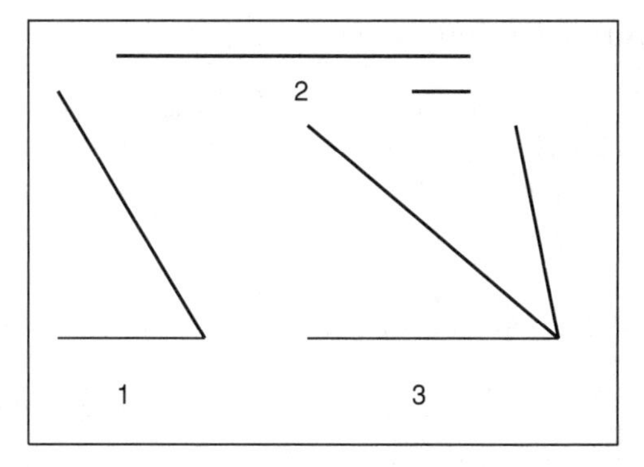

Fig. 8.2. Addition of the uncertainties a and b from Fig. 8.1: (**1**) The sum of the uncorrelated components is shown. (**2**) The two possible solutions for totally correlated components are presented. (**3**) Like in (**2**), but for partial correlation

Example. ***Addition of Uncertainties***

Combined uncertainties as shown in Fig. 8.2 are obtained by geometric addition of the uncertainty pairs shown in Fig. 8.1.

What is the meaning of the angles in Fig. 8.2? Because the correlation describes a relation between two uncertainty components it has no direction. For instance, the correlated and uncorrelated uncertainty components of a combined uncertainty are uncorrelated *to each other!* (Otherwise we could not add these components quadratically to give the total uncertainty, see Sect. 8.2.3.) This is why this angle (called ϕ in the previous discussion) has no further importance except that it would help us reconstruct the correlated and uncorrelated components from their sum.

8.1.1 Uncertainty of a Sum

Let the relation $F = x + y$ be given together with values x and y and their uncertainties Δx and Δy. To find the relation between ΔF and the uncertainties Δx and Δy we need to differentiate partially (as in Sect. 3.4):

$$\partial F/\partial x = 1\,, \quad \text{and} \quad \partial F/\partial y = 1\,.$$

For the case of *total correlation* we get the following via linear addition:

$$\Delta F = (\partial F/\partial x) \cdot \Delta x + (\partial F/\partial y) \cdot \Delta y = \Delta x + \Delta y\,. \tag{8.15}$$

For the case of *uncorrelated* uncertainties we get the following via quadratic addition:

$$\Delta F = \sqrt{(\partial F/\partial x)^2 \cdot (\Delta x)^2 + (\partial F/\partial y)^2 \cdot (\Delta y)^2} = \sqrt{(\Delta x)^2 + (\Delta y)^2}\,. \tag{8.16}$$

For both cases the following is true:

If a result is obtained by addition then the uncertainty components need to be added, too.

8.1.2 Uncertainty of a Difference

The relation $F = x - y$ is given together with values x and y and their uncertainties Δx and Δy. Again, we need to differentiate partially to find the relation between ΔF and the uncertainties Δx and Δy:

$$\partial F/\partial x = 1\,, \quad \text{and} \quad \partial F/\partial y = -1\,.$$

For the case of *total correlation* we get the following via linear addition:

$$\Delta F = |(\partial F/\partial x)\cdot\Delta x + (\partial F/\partial y)\cdot\Delta y| = |(1)\cdot\Delta x + (-1)\cdot\Delta y| = |\Delta x - \Delta y|\,, \tag{8.17}$$

which is $< \Delta x$ or $< \Delta y$.

Correlated uncertainties partially cancel each other because of the subtraction.

For *uncorrelated* uncertainties we get

$$\Delta F = \sqrt{(\partial F/\partial x)^2\cdot(\Delta x)^2 + (\partial F/\partial y)^2\cdot(\Delta y)^2} = \sqrt{(\Delta x)^2 + (\Delta y)^2}\,, \tag{8.18}$$

which is $> \Delta x$ and $> \Delta y$.

The second component becomes positive by squaring. Thus this combination of uncorrelated uncertainties gives a higher value than if they were correlated.

If uncertainty components that are uncorrelated (*independent*) are added or subtracted *the absolute uncertainties are added in quadrature.*

Examples

1. *Weighing.* A pressure gas container filled with propane weighed $m_t = 24.014\,\text{kg}$ at the time of delivery (when the container was still full). An empty container weighs $m_B = 13.563\,\text{kg}$ – this measurement value was obtained using the same scale. The calibration uncertainty of this scale for the range between 10 to 50 kg is given as ±0.05%.
   What is the *lower limit* for the detection of gas loss by weighing if a confidence level of 68% is sufficient?

$$\begin{aligned} m_G &= m_t - m_B\,,\\ \Delta m_t &= 0.0120\,\text{kg}\,,\\ \Delta m_B &= 0.0068\,\text{kg}\,,\\ \partial m_G/\partial m_t &= 1\,,\\ \partial m_G/\partial m_B &= -1\,,\\ \Delta m_G &= \Delta m_t\cdot\partial m_G/\partial m_t + \Delta m_B\cdot\partial m_G/\partial m_B\\ &= \Delta m_t - \Delta m_B = \pm 0.005\,\text{kg}\,. \end{aligned}$$

Note: We have limited ourselves to the calibration uncertainty that is inevitable. In addition it is obviously correlated. If these same uncertainties were uncorrelated, the uncertainty would be larger by a factor of 2.8.

2. *Buoyancy.* For precise mass measurements by weighing we need to consider that weight (not mass) is lost due to buoyancy in air (very much as in water) if the measurement is not conducted in a vacuum. Therefore, we need to correct for this weight loss (see also Sect. 7.3.2). If the measurement is conducted relative to a mass standard we need to acknowledge that this standard is also subject to buoyancy. The best estimate m for the mass is then determined from the measured mass value m_m as follows:

$$m = m_m + m_{aP} - m_{aN} ,$$

where m_{aP} and m_{aN} are the masses of the air displaced by the measurement sample and the standard, respectively.
The measured value be $m_m = (12.0000 \pm 0.0010)\,\mathrm{g}$, the volume of the sample be given as $V_P = 10.0\,\mathrm{cm}^3$. It is known that the mass standards are made of brass (with a density of $\rho_N = 8.4\,\mathrm{g/cm}^3$). Therefore we get a density of the sample of $\rho_P = (1.200 \pm 0.006)\,\mathrm{g/cm}^3$ based on the implicit uncertainty quotation of $\pm 0.5\%$ for the volume and a negligible uncertainty of the mass. So, including the air density at the time of the measurement of $\rho_L = (1.199 \pm 0.004)\,\mathrm{mg/cm}^3$, the best estimate m is gotten as

$$\begin{aligned} m &= m_m + \rho_L \cdot V_P - \rho_L \cdot V_P \cdot \rho_P/\rho_N \approx m_m + \rho_L \cdot V_P - m_m \cdot \rho_L/\rho_N \\ &= m_m \cdot (1 - \rho_L/\rho_N) + \rho_L \cdot V_P , \end{aligned} \tag{8.19}$$

with the following uncertainties:

$$\begin{aligned} \Delta m_m &= \pm 1.0\,\mathrm{mg} , \\ \Delta V_P &= \pm 0.05\,\mathrm{cm}^3 , \\ \Delta \rho_L &= \pm 0.004\mathrm{mg/cm}^3 , \\ \Delta \rho_N &= \pm 0.05\mathrm{mg/cm}^3 . \end{aligned}$$

The following dependences result by differentiating partially:

$$\begin{aligned} \partial m/\partial m_m &= 1 - \rho_L/\rho_N \\ \partial m/\partial V_P &= \rho_L \\ \partial m/\partial \rho_L &= V_P - m_m/\rho_N \\ \partial m/\partial \rho_N &= m_m \cdot \rho_L/\rho_N^2 . \end{aligned}$$

By applying the law of error propagation we get:

$$\Delta m = \tag{8.20}$$

$$\sqrt{(1-\rho_L/\rho_N)^2 \cdot (\Delta m_m)^2 + \rho_L^2 \cdot (\Delta V_P)^2 + (V_P - m_m/\rho_N)^2 \cdot (\Delta \rho_L)^2 + (m_m \cdot \rho_L/\rho_N^2)^2 \cdot (\Delta \rho_N)^2} .$$

Disregarding the quadrature, we have the following situation: The first component under the square root equals about 1 mg, the second one about 0.05 mg, the third one < 0.04 mg, and the fourth one is negligible, mainly due to $\rho_L \ll \rho_N$. Therefore the first component is the only dominant one, and we get

$$\Delta m \approx (1 - \rho_L/\rho_N) \cdot \Delta m_m \,, \quad \text{or in numbers,}$$
$$\Delta m = \pm 1.0\,\text{mg}\,.$$

8.1.3 Uncertainty of a Product

Let the relation $F = x \cdot y$ be given together with values x and y and their uncertainties Δx and Δy. By partial differentiation we get

$$\partial F/\partial x = y\,, \quad \text{and} \quad \partial F/\partial y = x\,.$$

For the case of *total correlation* between the uncertainties Δx and Δy the following is produced via linear addition:

$$\Delta F = (\partial F/\partial x) \cdot \Delta x + (\partial F/\partial y) \cdot \Delta y = y \cdot \Delta x + x \cdot \Delta y\,, \tag{8.21}$$

and after dividing by F,

$$\Delta F/F = \Delta x/x + \Delta y/y\,. \tag{8.22}$$

For the case of *uncorrelated* uncertainties we get the following via quadratic addition:

$$\begin{aligned} \Delta F &= \sqrt{(\partial F/\partial x)^2 \cdot (\Delta x)^2 + (\partial F/\partial y)^2 \cdot (\Delta y)^2} \\ &= \sqrt{y^2 \cdot (\Delta x)^2 + x^2 \cdot (\Delta y)^2}\,. \end{aligned} \tag{8.23}$$

After dividing by F the result is

$$\Delta F/F = \sqrt{(\Delta x/x)^2 + (\Delta y/y)^2}. \tag{8.24}$$

If a result is obtained by multiplication the *relative* uncertainties are *added.* In those cases where the uncertainties are uncorrelated (*independent*) the relative uncertainties need to be added *quadratically.*

Example. ***Electric Power in a Resistor***

The (direct current) power P at a certain operating point of some electronic one-port is defined by the current I and voltage V at that point

$$P = I \cdot V\,.$$

The values of I and V were determined to be

$$I = (3.500 \pm 0.011)\,\text{mA}\,, \quad \text{and}$$
$$V = (5.600 \pm 0.009)\,\text{V}\,.$$

Care was taken that the correlated part of the uncertainties (a common calibration uncertainty of ±0.1%) is not included in the uncertainties.

From above equation we obtain the relative (uncorrelated) uncertainty of P

$$\Delta P/P = \sqrt{(\Delta I/I)^2 + (\Delta V/V)^2} = \pm 0.35\%\,.$$

The calibration uncertainties of ±0.10% each – assumed to be 100% correlated – are linearly added to ±0.20%. By quadratic combination (Sect. 8.2.3) of this correlated with the uncorrelated uncertainty calculated above we get a total uncertainty of ±0.40%.

8.1.4 Uncertainty of a Ratio

The relation $F = x/y$ is given together with values x and y and their uncertainties Δx and Δy. To find the relation between ΔF and the uncertainties Δx and Δy, we need to differentiate partially (as in Sect. 3.4)

$$\partial F/\partial x = 1/y\,, \quad \text{and}$$
$$\partial F/\partial y = -x/y^2\,.$$

For the case of a *total correlation* we get the following via linear addition:

$$\Delta F = |(\partial F/\partial x)\cdot \Delta x + (\partial F/\partial y)\cdot \Delta y| = |\Delta x/y - \Delta y \cdot x/y^2|\,, \qquad (8.25)$$

and after dividing by F:

$$\Delta F/F = |\Delta x/x - \Delta y/y|\,. \qquad (8.26)$$

If these relative uncertainties are not only correlated, but also identical (identical calibration uncertainties, for example) the uncertainties fully cancel each other. This cancellation makes the quasi-absolute measurement method (Sect. 10.5.1) so potent.

If the uncertainties are *uncorrelated*, they need to be added quadratically:

$$\begin{aligned}\Delta F &= \sqrt{(\partial F/\partial x)^2 \cdot (\Delta x)^2 + (\partial F/\partial y)^2 \cdot (\Delta y)^2} \\ &= \sqrt{(\Delta x)^2/y^2 + x^2 \cdot (\Delta y)^2/y^4}\,. \qquad (8.27)\end{aligned}$$

After dividing by F the result is

$$\Delta F/F = \sqrt{(\Delta x/x)^2 + (\Delta y/y)^2}\,. \qquad (8.28)$$

If a result is a ratio of two components *their relative uncertainties are added quadratically*, but only in those cases where the uncertainties are uncorrelated (*independent*).

Example. ***Resistance of an Electrical Resistor***

The resistance R of a *linear* resistor can be obtained via the ratio of the operating point values voltage V and current I:

$$R = V/I\,. \tag{8.29}$$

The values of I and V have been measured with a digital multimeter, giving the following results:

$$I = (3.500 \pm 0.011)\,\mathrm{mA}\,, \quad \text{and}$$
$$V = (5.600 \pm 0.009)\,\mathrm{V}\,.$$

Care was taken that the correlated part of the uncertainties (a common calibration uncertainty of $\pm 0.10\%$) is not included in the uncertainties. So we get the relative uncertainty of R using (8.28)

$$\Delta R/R = \sqrt{(\Delta I/I)^2 + (\Delta V/V)^2} = \pm 0.35\%\,.$$

The calibration uncertainties of $\pm 0.10\%$ each – assumed to be 100% correlated – cancel each other. Thus the correlated uncertainty equals zero, and the total uncertainty is also $\pm 0.35\%$.

Note: Due to the great practical importance of the cancellation of correlated uncertainties in ratios (see quasi-absolute measurement method, Sect. 10.5.1) we give another, elementary proof that the above result is independent of the quality of calibration, and therefore is also independent of the calibration uncertainty.

Knowing that a current measurement is conducted indirectly via a voltage measurement we also know that current and voltage measurements conducted with the same instrument have the same calibration uncertainty, i.e., that they have an uncertainty component that is 100% correlated. If we extract this calibration factor c_f from these two measurement values we get

$$R = V/I = c_f \cdot V'/(c_f \cdot I') = V'/I'\,,$$

the common calibration factor cancels, and also its uncertainty. For the measurement of a linear resistance value via a current–voltage measurement the quality of the calibration of the measurement device is *absolutely irrelevant.*

8.1.5 Uncertainty of a Power (Root)

The relation between F and the value x is given by the equation $F = x^n$, together with a value of x, the parameter n, and the uncertainty Δx. To determine the relation between Δx and ΔF, we need to differentiate (as in Sect. 3.4):

$$\mathrm{d}F/\mathrm{d}x = n \cdot x^{n-1}\,. \tag{8.30}$$

So we get

$$\Delta F = (\mathrm{d}F/\mathrm{d}x) \cdot \Delta x = n \cdot x^{n-1} \cdot \Delta x\,. \tag{8.31}$$

After dividing by F the result is

$$\Delta F/F = n \cdot (\Delta x/x)\,. \tag{8.32}$$

If we chose a value of $n = 2$, for instance,

$$F = x^2 = x \cdot x\,.$$

The main difference from Sect. 8.1.3 is the fact that we are not dealing with a product of two uncorrelated quantities x and y, but with a product of identical quantities, i.e., their uncertainties are 100% correlated. Therefore, the uncertainties need to be added linearly. According to Sect. 8.1.3, this gives the following result:

$$\Delta F/F = (\Delta x/x) + (\Delta x/x) = 2 \cdot (\Delta x/x)\,,$$

just like using (8.32). As can be shown easily, this equation can be generalized for all positive integers n.

The above general equation can be applied to roots also, as roots can also be expressed as exponents: for instance, $\sqrt{x} = x^{1/2}$.

Problems

8.1. The relation $F = x^m \cdot y^n$, the values x and y, their uncertainties Δx and Δy, and the parameters m and n are given.

Determine (for fixed values of m and n):

(a) ΔF,
(b) $\Delta F/F$.

8.2. The mean diameter of a ball in a ball bearing has been determined with a caliper in various measurements to be (5.00 ± 0.05) mm.

(a) Why is it necessary to make multiple measurements? Calculate:
(b) the volume of the ball,
(c) its uncertainty.

8.1.6 Uncertainty of More Exotic Functions

As long as the mathematical function can be differentiated, it is possible to determine its dependence on its components by partial differentiation. Therefore, Sects. 8.1.1–8.1.5 do not contain new information; they are just meant as an exercise. In the following problem you can show how skilled you are in this field.

Problem

8.3. The phase shift between two sinusoidal signals of the same frequency can be measured with an oscilloscope by connecting one signal to the x-input terminal and the other to the y-input terminal. Then the (tilted) ellipse has to be evaluated – for 0 degrees it becomes a line, for 90 degrees a circle can be seen (these are Lissajous' figures). The sine of the phase shift φ can be determined from the ratio of the "half-width" x of the ellipse to the maximum extension X of it in the direction of the x-axis, $\sin\varphi = x/X$.

Calculate the uncertainty $\Delta\varphi$ from Δx and ΔX.

Note: Remember that the unit of the angle φ will be radians. From the result $\Delta\varphi = \tan(\varphi)\cdot\sqrt{(\Delta x/x)^2 + (\Delta X/X)^2}$ it looks as if for $x = 0$ (and consequently $\varphi = 0$) $\Delta\varphi = 0$ results. This would mean that we have a scientific result without uncertainty (Sect. 1.2). However, taking $\tan(\varphi) = \sin(\varphi)/\cos(\varphi)$ we get with $x = 0$ $\Delta\varphi = (\Delta x/X)/\cos(\varphi) \neq 0$.

8.2 Total Uncertainty

Usually (Chap. 4) it is implied that quantities having uncertainties are part of a normal distribution (Sect. 5.2.3). For practical purposes it suffices that this requirement is fulfilled sufficiently well with respect to the dominant uncertainty components (Sect. 6.2.3).

8.2.1 Adding Correlated to Uncorrelated Uncertainties

It is general practice to combine random and systematic errors in quadrature (Sect. 3.4) to yield total uncertainties. This is in accord with the following remark taken from a monograph of the OECD (Smith, DL (1991) *Probability, Statistics, and Data Uncertainties in Nuclear Science and Technology*, American Nuclear Society, LaGrange Park): "There is no evidence that you cannot treat random errors and systematic errors the same way." Some even do so if the systematic "errors" are deviations (with a sign) and not uncertainties. This, however, is utter nonsense!

The above remark should be self-evident by now because we know (Sects. 7.2.2 and 7.2.3) that correlated and uncorrelated uncertainties are basically the same. It hinges solely on whether uncertainties are independent of each other to have them added in quadrature or not. So we can summarize our expectations:

- Uncorrelated uncertainties must be combined in quadrature to yield combined uncorrelated uncertainties.
- Combined uncertainties that are independent of each other must be combined in quadrature even if some (or all) components contain (properly added) correlated uncertainties.

- The total uncertainty of a single best estimate can be obtained by adding the combined uncorrelated uncertainties and the combined correlated uncertainty components in quadrature because these two are obviously independent of each other.

8.2.2 Total Uncertainty of a Single Best Estimate

Whether a best estimate is given by a single value or by just one parameter (like the mean value of a data set) it is always appropriate to add all uncertainties to yield the total uncertainty of this best estimate. Correlated uncertainty components of single data are combined linearly according to the general law of error propagation (see the beginning of this chapter). If done correctly, the combined correlated uncertainty components will not be correlated to other uncertainty contributions. Therefore, they must be treated as independent (i.e., uncorrelated) uncertainties and must be combined with other uncertainty components by addition in quadrature (error propagation of independent uncertainties, see also Sect. 3.4).

In the case of a *single data value* uncertainties of measurement parameters that affect the scale only indirectly (e.g., ambient temperature, air pressure, angle or projectile energy in a cross section measurement) are, as a rule, not correlated to any other uncertainty so that they are independent of other uncertainty contributions. Therefore, the uncertainty of a measurement parameter must be added in quadrature, after the effect of this parameter on the data value has been established (e.g., Sect. 7.3.2, effect of temperature on the forward voltage of a semiconductor diode). That is, such uncertainty contributions, like any other, must be treated as uncorrelated uncertainties.

8.2.3 Total Uncertainty of Data Sets

Before best estimates are extracted from data sets by way of a regression analysis, the uncertainties of the individual data values must be determined. In this case care must be taken to recognize which uncertainty components are common to all the values, i.e., those that are correlated (systematic).

The uncorrelated uncertainty contributions of the individual data points are independent both of each other and also of the systematic uncertainty contributions of the data set. Thus they can be added quadratically to give a total uncorrelated uncertainty. In most cases, it will be prudent not to merge the combined correlated with the combined uncorrelated uncertainties to yield the total uncertainty of each individual data value. This is demonstrated in Fig. 8.3, where the best estimate, a regression line, is compared with the individual data values from which it is derived. It is just wrong to include correlated (e.g., scale) uncertainties in the uncertainty bars of a scatter plot because a scale uncertainty has no influence on the shape of the data dependence. (In Sect. 4.2.1 we have reduced the degrees of freedom by $f = 1$ because the scale of the mean and the scale of the data necessarily agree.

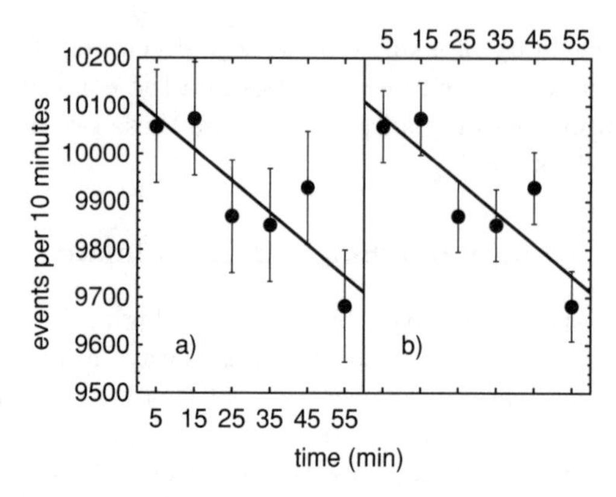

Fig. 8.3. Data points of a data set with line of regression. The *uncertainty bars* in the *left portion* include the scale uncertainty, those on the *right* do not

Therefore the scale factor is exactly one, and consequently no scale uncertainty exists!) Thus, *for data sets the scale uncertainty should not be included into individual uncertainty values but given separately* (see, e.g., Sect. 7.5).

Note : As shown in Sect. 7.4.2, correlated uncertainty components do not change the values of the parameters of the line as obtained by linear regression *if all data values have the same weight*. However, the uncertainties of the parameters depend on the size of the uncertainly!

In the left portion of Fig. 8.3 data values with uncertainty bars that include the scale uncertainty are given together with the regression line; in the right portion the uncertainty bars contain uncorrelated uncertainties only. Obviously, in case (a) the data are overfitted, whereas in case (b) the line cuts just two thirds of the uncertainty bars, as required for a 1σ confidence level .

Systematic Uncertainties of Measurement Parameters

If individual data values of a data set are not independent of each other, some of their uncertainty components will be correlated (systematic). If these systematic components are independent of each other they may be combined quadratically to a total systematic (scale) uncertainty.

Systematic uncertainties that are *not* scale uncertainties, e.g., uncertainties in the parameters of a measurement, must be handled differently. If the effect of a parameter change is the same for each data value then the corresponding uncertainty can be taken into account as a contribution to the scale uncertainty. This contribution to the scale uncertainty will, in general, be independent of other components of the scale uncertainty. Therefore, it is uncorrelated to the other components and *must be added in quadrature* to the rest of the scale uncertainty. If their effect on the scale is *not* the same for each data point, then it cannot be included in the scale uncertainty and such systematic uncertainties *must be stated separately.*

8.3 Using Internal Uncertainties for Diagnosis

8.3.1 Reliability of the Space Shuttle

When Richard P. Feynman tried to estimate the reliability of the space shuttle (after the Challenger accident in the year 1986), his approach was quite similar as with uncertainties (as described in the book Feynman RP (1992) *What Do You Care What Other People Think?* Bantam Dell, New York).

The management at NASA induced a failure rate of 10^{-5}, i.e., *one* accident in 300 years, with *one start each day*. In reality, this number has to be considered wishful thinking (as cited by Feynman: "The probability of success of manned space travel necessarily has to be close to 1", namely 0.99999) rather than the result of a realistic, reproducible probability calculation. This calculation would not have been possible with the data material available after only a couple of dozen of previously conducted successful starts (inductive method).

With the insider knowledge of the engineers, considering the reliability of the propulsion system a failure rate of about 1 in 200 was deduced, i.e., a success probability of 0.995. Only such internal properties of the space shuttle (corresponding to the *internal* uncertainties) can be used sensibly to judge its reliability before accidents happen. Trying to induce the failure rate from a (greater) number of accidents (i.e., determining the external uncertainty) would be nonsense. It would be not only wearisome, but above all, such a procedure would require that no technical improvements of the space shuttle have occurred during the time of observation (i.e., all data values must have the same weight).

8.3.2 Analogy to Bayes' Principle

When evaluating experiments we encounter a similar dilemma between the *deductive* prediction using established relations (comparable to *internal* uncertainties) and the *inductive* procedure using the distribution of the measurement values (comparable to *external* uncertainties). A generalization of the inductive inference is given by Bayes' Principle; it is of great help when wanting to draw conclusions from the measurement results as to their causes.

Example. ***Comparing the Deductive and Inductive Methods in a Game of Dice***

If we know the property of a die (e.g., that it is "not loaded", i.e., fair) we can predict the probability for a certain throw or any combination of throws (deductive procedure). If, on the other hand, we have combinations of throws, a statement (with uncertainty) on the quality of the die about possible irregularities can be induced, e.g., by applying Bayes' Principle on the results of this "measurement". A comprehensive introduction to Bayes' Principle can be found, for instance, in the book Sivia DS (1997) *Data Analysis. A Bayesian Tutorial*. Oxford University Press, Oxford.

8.3.3 chi-Squared Test

In error analysis the so-called "*chi*-squared" is a measure of the agreement between the *uncorrelated* internal and the external uncertainties of a measured functional relation. The simplest such relation would be time independence. Theory of the chi-squared requires that the uncertainties be normally distributed. Nevertheless, it was found that the test can be applied to most probability distributions encountered in practice.

If everything is right, the external uncertainties – represented by the deviations $(m_i - y_i)$ – and the internal uncertainties Δy_i are of the same size *on average*. That is, their ratio, when expressed in the following *reduced* form, equals one:

$$\chi^2 = \frac{1}{n-f} \cdot \sum_{i=1}^{n} \left(\frac{m_i - y_i}{\Delta y_i} \right)^2 \approx 1 . \tag{8.33}$$

As usual, n stands for the number of data points, $(n - f)$ for the number of degrees of freedom, m_i for the corresponding best estimate (i.e., the functional value) at the position of data point i, y_i for the corresponding data value, and Δy_i for the corresponding (uncorrelated) internal uncertainty. *No less than ten degrees of freedom should be available to make the result of the test trustworthy.*

For the limiting case of an infinite number of data points and under the assumptions

- that the functional relation (the "theory") is correct,
- that the internal uncertainties have been determined correctly (i.e., that they are uncorrelated and have the correct value), and
- that the data values are normally distributed

chi^2 *must be* 1.

When *chi*^{2} obviously differs from 1, the following conclusions can be made:

1. Under the assumption that the *theory is correct*, the following is true:
   - For *chi*$^2 < 1$ the internal uncertainties are too large; they might contain correlated (systematic) components.
   - For *chi*$^2 > 1$ contributions to the uncertainties have been overlooked.
2. Under the assumption that the *uncertainties have been determined correctly*, the following is true:
   - For *chi*$^2 < 1$ the functional representation of the data uses more parameters than supported by the data – this is called "overfitting". Obviously *chi*^{2} becomes zero if the number of free parameters is the same as the number of the (independent) data values.
   - For *chi*$^2 > 1$ a function of a higher degree than used is required for the representation of the data values.

With a finite number of degrees of freedom chi^2 will deviate from 1 even if everything is correct. The more degrees of freedom are involved the more significance a deviation from one will have. *With less than ten degrees of freedom no reliable result can be expected.*

Example. *Comparison of two Fits*

In Fig. 4.3 (Sect. 4.2.3) two choices of a best estimate are given for the 12 data values from Table 6.2; on the one hand, via the mean value (a line with a slope of zero), and on the other hand, by linear regression.

1. For the first case (radiation intensity does not vary in time) we obtain
$$\chi^2 = \frac{1}{11} \cdot \frac{1}{9910.1} \cdot 110219 = 1.011 .$$
The answer is very close to 1, meaning that the dispersion of the data (as a measure for the external uncertainty) and their internal uncertainty $\pm\sqrt{y_m}$ agree very well.
2. In the case of the linear regression that results in a line with negative slope we get $chi^2 = 0.85$.
From these two answers we conclude that the first solution is better – its chi^2 is closer to 1. In the second case we have overfitting, i.e., this function is of too high a degree. Thus it has been shown that it is more likely that the count rate is constant in time than decaying with a time constant as obtained by linear regression.

Note:

- The half-life of the source in Sect. 4.2.1 equals 43 years. This corresponds to a slope that has a much smaller absolute value than the one determined by linear regression. (In Fig. 4.3 the decrease during the 12 min shown in the graph would be much smaller than the thickness of the horizontal line!) Such a slope would give a chi^2 minimally closer to one than that obtained for the mean. However, this slope cannot be deduced from the data because of their insufficient precision.
- For the case of the linear regression we have tacitly assumed that the uncorrelated uncertainty of the time measurement is negligible. If this is not the case, these uncertainties and Δy_i need to be combined – as described in Sect. 9.1.4 – before applying the chi-squared test.

Goodness of Fit

The (reduced) chi-squared test is frequently used as criterion when *"fitting"* data with a computer – for the decision of which degree the polynomials need to be to represent the data best. As we have seen, *chi*-squared will not be too far from 1 if, on the one hand, the polynomials to be fitted have the correct degree, and, on the other hand, if

Table 8.1. Record of time-independent data

Clock time	Data value	Uncertainty
t_i	y_i	Δy_i
2:10:00.00	9974	118.2
2:20:00.00	10023	118.4
2:30:00.00	9852	117.7
2:40:00.00	9868	117.8
2:50:00.00	9979	118.2
3:00:00.00	9765	117.3

- the uncertainties have the correct size,
- a sufficient number of degrees of freedom (at least 10) are available.

The uncertainty of the parameters of such a fit is usually given by estimates, called *asymptotic standard errors*, rather than by confidence intervals. These are attained from the variance–covariance matrix after the final iteration. For linear fits these asymptotic standard errors are the standard deviations because they are derived like the standard deviations of the parameters in a linear least-squares problem (Sect. 6.3.2). For nonlinear fits they tend to be overoptimistic, and merely give some qualitative information (for exceptions see below).

The correlation between the parameters in the region of the solution can be obtained from their correlation matrix. This correlation determines whether a change in *chi*-squared due to a change in one parameter can be counterbalanced by changing some other parameter. If this is the case there will be an off-diagonal matrix element with a value close to 1. If its sign is positive, the corresponding two parameters act in the same direction, otherwise they act in opposite directions. If all parameters are independent of each other, *only* the main diagonal matrix elements will not be zero. In such cases, the standard deviations of the parameters can be obtained from the asymptotic standard errors also for nonlinear fits.

Problem

8.4. For the data values listed in Table 8.1 theory requires that the functional relation is a horizontal line.

(a) Make a scatter plot (including uncertainty bars, see Sect. 9.1.3) and draw the horizontal line, the best estimate, as obtained by the arithmetic mean.
(b) How can we tell "by taking a closer look" at the graph that the uncertainties are (partially) correlated?
(c) Support the above "notion" by determining *chi*-squared.
(d) Under the (arbitrary) assumption that the straight line is the right function, we can subtract a correlated component (of equal size) from each

uncertainty (Sect. 3.4.2) so that $chi^2 \approx 1$ results. Do this, at least approximately.

(e) Determine *chi*-squared after subtracting the correlated component.

(f) Why must the above results of the chi-squared test be treated with some caution?

9

Presentation and Estimation of Uncertainties

Any scientific data without (a stated) uncertainty is of no avail. Therefore the analysis and description of uncertainty are almost as important as those of the data value itself. It should be clear that the uncertainty itself also has an uncertainty – due to its nature as a scientific quantity – and so on. The uncertainty of an uncertainty is generally not determined.

9.1 Graphic Presentation, Also of Uncertainties

Great amounts of data are better visualized graphically than, for instance, by data tables. The tools nowadays available (computers) allow for much more intricate graphic presentation than necessary, or even at times appropriate or helpful. The required integrity of science should not allow that data of possibly low(er) quality be camouflaged by mazy graphic presentation. Unluckily, this is not always the case – often it seems that the wrapping has become more important than the contents, even in scientific presentations. I can find no other reasons for those colorful exhibitions enabled by the use of computers. Nothing is necessarily a good or reasonable thing to do, just because it can be done, nor can the use of a computer guarantee that the data handled by it are any good.

9.1.1 Basics

Graphical presentation is used to visualize the functional relation between two or more quantities. The following points should generally be observed (when using an orthogonal coordinate system):

- Correct labeling of the axes, including the quantity presented and its units.
- The text should remain legible, even if the editor reduces the size of the diagram.

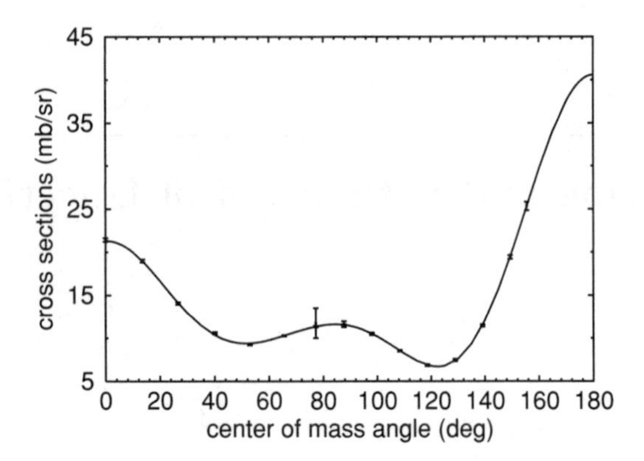

Fig. 9.1. Linear diagram of the data of Table 10.2. Note that the origin is suppressed

- Use as much of the area available inside the diagram for the presentation of the data (e.g., optimum scaling of the coordinates, or zero point suppression, e.g., as done in Fig. 9.1).
- Transformation of the variables (and the corresponding axes, e.g., as done in Fig. 10.1) to establish a simpler relation (e.g., taking logarithms to linearize an exponential relation as done in Fig. 9.3 and detailed in Sect. 9.1.5).
- Transformation of one or both variables to show the difference between the data and the regression curve over the *whole* range (logarithmic presentation if a large difference exists between the smallest and the largest data value, as in Fig. 9.2, or zero point suppression that *must be indicated* – if not obvious).
- Self-explanatory captions under the diagrams.

For a diagram with a nonorthogonal coordinate system the above is true correspondingly.

Comparing Fig. 9.1 with Fig. 9.2, the advantages of the logarithmic presentation are easily recognized. In Fig. 9.1 the data from Table 10.2 are shown in a linear presentation (with suppressed zero point), and in Fig. 9.2 they are shown in a semilogarithmic presentation. Most data (75%) lie between 6 and 12 mb/sr. In the linear case only 13% of the area of the diagram is designated for their presentation. By suppressing the zero point this percentage is raised to 15%. In the semilogarithmic presentation this percentage is increased to 37%, an increase of the resolution for most of the data by a factor of $2^1/_2$! In Table 10.2 the uncertainties are given in percent, also indicating that a logarithmic presentation should be used; this way the bars for uncertainties of the same percentage have the same length, independent of their position in the graph.

It is important to pay heed to the following detail: a disadvantage of logarithmic diagrams is that a graphical integration is not possible, i.e., the area under the curve (the integral) is of no relevance.

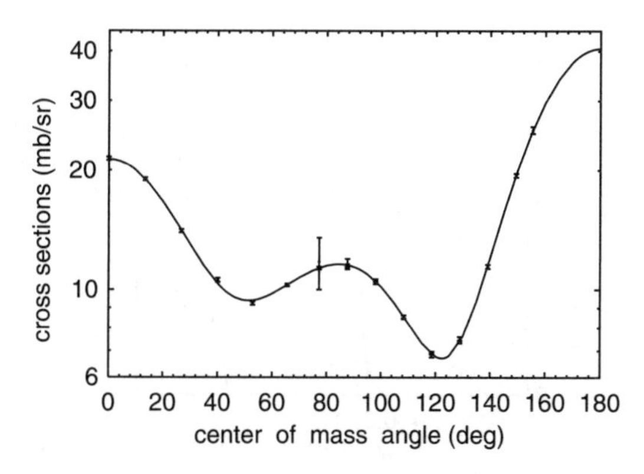

Fig. 9.2. Semilogarithmic diagram of the data of Table 10.2

9.1.2 Charts

Nonorthogonal diagram types like the circular or pie chart or the stem and leaf chart are not suited for data with uncertainties. Even some orthogonal types cannot be used, for instance, bar charts like the one in Fig. 5.4, because uncertainties are generally presented as uncertainty bars (Sect. 9.1.3) that emanate from a point. Thus it is not possible to use uncertainty bars in connection with bar charts.

Data with uncertainties are best presented via

- scatter plots (or scatter diagrams, e.g., Figs. 4.3 and 8.3),
- polygon charts (e.g., Fig. 5.2),
- histograms (e.g., Fig. 5.2).

When using scatter diagrams it is important not to connect the dots with (straight) lines. These lines between scattering points would only be misleading and are definitely meaningless.

9.1.3 Uncertainty Bars (Error Bars)

In orthogonal coordinate systems the uncertainty of a data value is given by an error bar that is parallel to the corresponding axis, as can be seen in Figs. 8.3 and 9.2. Typically, these are 1σ uncertainties, where the corresponding data point lies in the middle of the interval. In few cases (e.g., Fig. 9.3) uncertainty bars are shown in parallel to both axes. In most cases just the uncertainty of the dependent variable is shown.

9.1.4 Uncertainty Rectangle (Error Rectangle)

If a data point has uncertainty bars both in the y- and the x-direction (Fig. 9.3) these uncertainty bars define an error rectangle. Such *uncertainty rectangles*

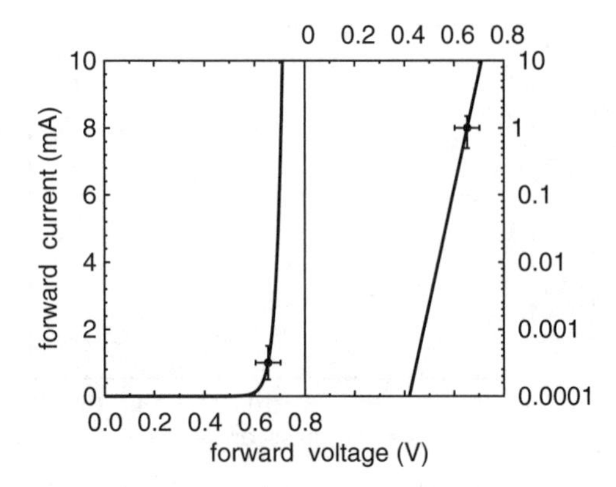

Fig. 9.3. The characteristics of a planar silicon diode presented linearly and semilogarithmically. The operating point corresponds to the one in Example 3 in Sect. 7.3.2. The x-uncertainty bar and the y-uncertainty bar together give the uncertainty rectangle

are useful for visualization, but are rather inconvenient for further numerical evaluation. In those cases where the functional relation between the two variables is known for the proximity of the data value, the uncertainty rectangle can be transformed into a single uncertainty bar. If the uncertainties are independent of each other, i.e., if they are uncorrelated, this is done by adding the two uncertainty contributions in quadrature (Sect. 3.4).

Problem

9.1. Strictly speaking, in Example 3 of Sect. 7.3.2, the triplet of uncertainty bars form an *uncertainty volume (cuboid)* representing the three uncertainties assigned to the data value. The operating point of a silicon semiconductor diode at (22 ± 3)°C be given by a current of $(1.000 \pm 0.010)\,\mathrm{mA}$ and a forward voltage of $(0.653 \pm 0.004)\,\mathrm{V}$. The temperature dependence of this voltage is $-2\,\mathrm{mV/°C}$, and the diode current increases exponentially with the voltage. From semiconductor theory we get the dependence of the voltage on the current (i.e., the impedance Z of the diode at the operating point) as $0.99\,\mathrm{V/A}$ $(= \Omega)$.

Combine the three uncertainty components of the forward voltage at the measured current of 1.000 mA and the temperature of 22°C.

9.1.5 Linearization

Semilogarithmic or double-logarithmic presentations are often used to linearize relations that are nonlinear. The exponential function

$$y = a \cdot \mathrm{e}^{bx} \tag{9.1}$$

can thus be transformed into

$$\ln(y) = b \cdot x + \ln(a) . \tag{9.2}$$

The ordinate is now logarithmic, and the abscissa is linear. This is called a semilogarithmic presentation. In Fig. 9.3 this type of diagram (i.e., a semilogarithmic plot) is compared with the linear presentation of the characteristics of a diode.

Taking the logarithm of the power law

$$y = a \cdot x^b \tag{9.3}$$

yields the linear equation

$$\log(y) = b \cdot \log(x) + \log(a) . \tag{9.4}$$

In this case we choose the double-logarithmic presentation, i.e., the tick marks on both axes are distributed logarithmically. In such a presentation the slope of the line gives the exponent of the power law.

Note: In the logarithmic presentation the data value is not in the center of the uncertainty bar.

9.2 Correct Presentation of Uncertainties

As it is not usual that uncertainties are stated with *their* uncertainties, the number of figures used in their quotation should reflect this uncertainty (see implicit presentation of uncertainties, Sect. 3.2.4). As discussed there, it should be avoided that rounding introduces an uncertainty that distorts the actual accuracy of an uncertainty. If you make a statement on an uncertainty like ±0.1, it is clear that the rounding uncertainty of the uncertainty is ±50%. Such a situation must definitely be avoided. On the other hand, it does not make sense to state the value of an uncertainty more accurately than its data value. A good rule is to determine an uncertainty in such a way that one additional decimal digit (when compared to the data value) is necessary for the (implicit) presentation of it. If done so the data value will have one additional decimal digit, which is insignificant.

This rule for stating uncertainties is not necessarily in accord with other recipes found elsewhere. However, it is adhered to in practical life (e.g. Audi G, Wapstra AH, and Thibault C (2003) *International Atomic Mass Table*. Nuclear Physics A729: 337). Looking up, e.g., the mass of the neutron one finds (1,008,664.91574±0.00056) micro-mass units. Obviously, the last decimal digit of the mass value is *not* significant.

Examples. ***Correct Data Presentation***

The following presentations are *correct*:

- 1.10 ± 0.10; the data value is known $\pm 9\%$ ($0.10/1.10 = 0.09$), the uncertainty within $\pm 5\%$ ($0.005/0.10 = 0.05$) (Keep in mind that it is vital not to drop the "0" in the last digit!),
- 9.89 ± 0.10; the data value is known $\pm 1.0\%$, the uncertainty within $\pm 5\%$,
- 3.00 ± 0.30; the data value is known $\pm 10.0\%$, the uncertainty within $\pm 2\%$,
- 3.0 ± 0.3; the data value is known $\pm 10\%$, the uncertainty within $\pm 17\%$ (this limiting case should rather be avoided).

The following presentations are *not* correct:

- 1.1 ± 0.1; the data value is known $\pm 9\%$, the uncertainty within $\pm 50\%$ (the maximum rounding uncertainty of ± 0.05 corresponds to 50% of the uncertainty, or 4.5% of the data value having an accuracy of $\pm 9\%$, respectively.)
- 1.1 ± 0.10; the data value does not have a sufficient number of digits,
- 1.104 ± 0.104; the data value is known $\pm 9\%$, the uncertainty is given within $\pm 0.5\%$ (The last two digits of the data value are not significant; therefore, it does not make sense to state more than two digits after the floating point.)

Problem

9.2. How much shorter could a counting experiment have been if the result is given by 1.1 ± 0.1 instead of 1.07 ± 0.05 without losing precision? (Those who need inspiration should see Sect. 10.1.1.)

9.3 Finding the Size of Internal Uncertainties

An accurate determination of the uncertainty can be more bothersome than that of the data value. Luckily, the "quadratic" addition of individual uncorrelated uncertainty components when combining them ("law of error propagation", see Sect. 3.4) is of great help – only the dominant uncertainty components matter. It is sufficient to show that the other components are small compared to the largest dominant one. For this we just need an estimate of the upper limit of such uncertainties to determine whether it is legitimate to disregard these components.

Example. ***Dominant Uncertainty Contribution***

In count rate measurements of nuclear radiation the (main) contributions to the uncertainty stem from:

- counting statistics (due to the random nature of radioactive decay of nuclei),
- time measurement,
- dead-time correction.

Under the assumptions:

- that $N = 10{,}000$ events were recorded in a time interval t,
- that the dead time t_D amounts to 2.0%,
- that the time t has been measured with the help of a quartz-based timer,

the *dominant uncertainty* is that of the counting statistics ($\pm 0.99\%$, after dead time correction). The typical accuracy of simple quartz-based timers is $< 0.01\%$, and the accuracy of the dead time correction is estimated to be smaller than 10% of its value (i.e. $<0.2\%$). From the equation for the event rate ER:

$$ER = N/(t - t_D) = N/(0.98 \cdot t)\,, \tag{9.5}$$

we obtain the relative uncertainty components of N as $\pm 0.99\%$, of t as $< 0.01\%$, and of the dead-time factor $< (0.20/0.98)\%$. The resultant total (percentage) uncertainty $\Delta ER/ER$ becomes $< 1.01\%$, rounded to $\pm 1.0\%$. There is hardly any difference between the calculated total uncertainty of $\pm 1.01\%$ and its dominant component (statistical uncertainty, $\pm 0.99\%$); therefore an accurate determination of the other two components is neither necessary nor sensible.

Problem

9.3. Verify the calculation of the uncertainties given in the example of Sect. 9.3.

9.3.1 Ideal Situation

Ideally all measurements should be based on the primary standards (Sect. 2.1). Then the uncertainties are given by the accuracy of "remeasuring" these standards and of how these basic quantities and the actual measurement value relate to each other. The simple example of a mass determination shall demonstrate that it is hard if not impossible to follow such a practice: Imagine using the kilogram prototype (which is kept in Sèvres near Paris) for every mass determination! Even if a national bureau of standards exists close by, one would hardly use their (derived) standards for a mass determination.

9.3.2 Pragmatic Solution

In the real world we can assume that the manufacturer of a measuring device has tied the scale of this device to the standards available from the national bureau of standards. The accuracy of this calibration procedure is documented in the specifications of the instrument.

A case is simple for the experimenter if

1. The measurement uncertainty is described sufficiently well in the specifications of the instruments in use, although in many cases this will not be a 1σ uncertainty.
2. As in the above discussed case of counting radioactive events, a proven and established theory for the determination of the dominant uncertainty exists.

In all other cases we need to come up with some idea of how to gain information on the accuracy of the measurement setup, e.g.,

3. By remeasuring a data point of a very well known value, a secondary "standard", to calibrate the apparatus.
4. By "estimating the uncertainties".

If a known data value (naturally, of the same type as the unknown value we want to measure) is measured with the identical apparatus, we can derive the intrinsic accuracy of the measurement setup from the knowledge of the measured value and of the known (standard) value.

Note: Interpreting the new measurement as a measurement relative to the measurement of the known reference value makes a lot of sense, as many uncertainty components are strongly correlated in such a comparison, and these uncertainties cancel because we are dealing with ratios (Sect. 8.1.4). The other possible option is using the measurement of the known value for calibrating the instrument, i.e., the known data value is treated as secondary standard (Sect. 10.5.1).

9.4 Estimating the Size of (Internal) Uncertainties

As uncertainties of scientific data values are nearly as important as the data values themselves, it is usually not acceptable that a best estimate is only accompanied by an estimated uncertainty. Therefore, only the size of nondominant uncertainties should be estimated. For estimating the size of a nondominant uncertainty we need to find its upper limit, i.e., we want to be as sure as possible that the uncertainty does not exceed a certain value.

Estimated uncertainties are definitely not part of a normal distribution because they are unavoidably subjective. The same can also be true when extracting data with uncertainties from literature or when using instrument specifications taken from manuals. However, as pointed out before (Sect. 6.2.3), it does not matter that we are not dealing with a normal distribution as long as this applies to *nondominant* uncertainty contributions. This is even more true if these contributions are so small that they may be ignored entirely.

Even though there is a generally applicable method of finding *lower limits* for uncertainties, *estimating their upper limits* is a rather subjective procedure.

9.4.1 Finding Upper Limits

The only good reason for estimating upper uncertainty limits in the course of a serious uncertainty analysis is to prove that the corresponding uncertainties may be ignored, i.e., that they are smaller by at least a factor of 3 than the largest dominant uncertainty component. If the estimated uncertainty component is one of the dominant ones, an alternative reliable way of determining this uncertainty component should be explored. If none is found, the final uncertainty will be something like a maximum uncertainty (Sect. 3.2.5).

Examples

1. *Dependence on the Ambient Temperature.* The experimenter knows from experience that the temperature in her lab varies about $\pm 3°$C. From some source (ideally from one of her own measurements) she knows the temperature coefficient $\theta\%/°$C of her measurement value. Without actually recording the temperature she can determine the contribution of the temperature change to the uncertainty to be $\pm < 3\theta\%$. If this contribution cannot be ignored, either the lab should be air-conditioned or the temperature during the measurement must be recorded as another measurement parameter.
2. *Accuracy of the Measurement Time.* In the example in Sect. 9.3. the uncertainty of the measurement time was estimated using the knowledge that the time base of the clock is controlled by quartz keeping the frequency constant within at least $\pm 10^{-4}$. If this estimated time uncertainty is too large to be ignored, we either need to calibrate the clock with a clock of higher accuracy or we need to use this other clock itself for the experiment.
3. *Uncertainty of a Correction.* Corrections should be small in the first place. Then it would not be necessary to determine their uncertainties because these uncertainties will *not* be dominant. For such cases, assuming an uncertainty of the correction of $< 50\%$ should be on the safe side. If the uncertainty of a correction cannot be determined and is expected to be $> 50\%$, a better measurement procedure that does not necessitate such little-understood corrections should be found. In the example in Sect. 9.3, a – very safe – $< 10\%$ uncertainty was assumed because the procedure for dead time correction in the instrument is well established; see also Sect. 4.1.4.
4. *Mass Determination* (example in Sect. 8.1.2, same symbols and same basic equations as there). Considering the buoyancy in air, the best estimate m is obtained from the measurement value m_m as

$$m = m_m + m_{aP} - m_{aN} .$$

Here m_{aP} and m_{aN} are the masses of the air displaced by the measurement sample and mass standard respectively. Limiting ourselves only to measurable quantities we obtain (8.19) and (8.20)

$$m = m_m + \rho_L \cdot V_P - \rho_L \cdot V_P \cdot \rho_P/\rho_N \approx m_m + \rho_L \cdot V_P - m_m \cdot \rho_L/\rho_N$$
$$= m_m \cdot (1 - \rho_L/\rho_N) + \rho_L \cdot V_P , \quad (8.19)$$

and

$$\Delta m = \quad (8.20)$$
$$\sqrt{(1-\rho_L/\rho_N)^2 \cdot (\Delta m_m)^2 + \rho_L^2 \cdot (\Delta V_P)^2 + (V_P - m_m/\rho_N)^2 \cdot (\Delta\rho_L)^2 + (m_m \cdot \rho_L/\rho_N^2)^2 \cdot (\Delta\rho_N)^2} .$$

If the uncertainty of the mass measurement is $\Delta m_m = \pm 1.0\,\text{mg}$, how small must the other three uncertainty components be so that they can be disregarded?

- From $\rho_L \cdot \Delta V_P < 0.33\,\text{mg}$ we get (using the values from Sect. 8.1.2) $\Delta V_P < 0.33/1.199\,\text{cm}^3$, and $\Delta V_P/V_P < 2.75\%$;
- from $(V_P - m_m/\rho_N) \cdot \Delta\rho_L < 0.33\,\text{mg}$ we get $\Delta\rho_L < 0.039\,\text{mg/cm}^3$, and $\Delta\rho_L/\rho_L < 3.2\%$;
- and from $(m_m \cdot \rho_L/\rho_N^2) \cdot \Delta\rho_N < 0.33\,\text{mg}$ we obtain $\Delta\rho_N < 1.6\,\text{g/cm}^3$, and $\Delta\rho_N/\rho_N < 19.4\%$.

For each of these three uncertainty components it is easy to estimate (with great certainty) that they are smaller than their calculated upper limit; thus they can be disregarded.

If the specifications concerning the accuracy of an instrument (provided by the manufacturer) have been lost, the values listed in Table 9.1 might be helpful when trying to estimate uncertainties.

The uncertainty values obtained this way are only good for showing that the corresponding uncertainty is negligible. If this is not the case, the values are **not** to be used!

Problems

9.4.

(a) How accurate must a watch with a battery at least be so that it is not necessary to adjust the time during the lifetime of the battery (2 years)? A maximum of one minute gain or loss is allowed during this time span.

(b) For those who like to split hairs: Which "environmental parameter" would need to be specified to make the accuracy requirement scientifically sound?

9.5. The rule that dominant uncertainty components should not be estimated was broken in this book at least once. Where?

Table 9.1. Typical, but not maximum uncertainties of various types of measurements. These values can *only* be used to show that the corresponding uncertainty component may be disregarded in case of an addition in quadrature. These numbers are under no circumstances suited for being part of the final uncertainty value

	Typical uncertainty value
Length measurement:	
Ruler with millimeter marks	0.2 mm
Caliper	0.1 mm
Digital caliper	0.02 mm
Micrometer	0.02 mm
Mass measurement:	
Precision weights > 100 g	< 3 mg
Precision weights > 1 g	< 0.5 mg
Precision weights > 0.1 g	< 0.1 mg
Precision weights > 0.02 g	$< 0.2\%$
Precision weights > 0.5 mg	$< 6\%$
Time measurement:	
Mechanical stop watch	< 0.2 s
Electronically triggered quartz-based clock	< 0.1 ms
Temperature measurement:	
Laboratory-type liquid thermometer	< 0.5 K
Electrical measurements (relative to full scale):	
Analog instruments (quality factor)	0.1 ... 5%
Digital instruments	$< 0.1\%$

9.4.2 Finding Lower Limits

There is one sure way to find the *lower limit* of an uncertainty: The uncorrelated uncertainty of an arithmetic mean value is greater than (or possibly equal to) the external uncertainty determined from the mean variation of the time series. In other words: The precision of a measurement can never be better than its reproducibility (Sects. 4.2.2 and 6.2.5).

10

Feedback of Uncertainties on Experiment Design

10.1 Optimizing Experiments

There is a saying that an experiment can only be done optimally when one knows the results – this saying contains some truth. However, ample experimental experience, especially in the field of the experiment to be conducted can help performing an experiment of very good quality at the first attempt.

A (short) *pre-experiment* might be necessary if we are covering new ground with the experiment and if larger machines (like accelerators) with limited machine time available are employed. Such a pre-experiment not only shows which areas of the experiment are especially prone to difficulties, but an analysis of the uncertainty components enables us to find out which components of the experiment must be improved to increase the overall accuracy. From what we have already discussed in Sect. 3.4.1 it should be clear:

> *Only* by improving the dominant uncertainties can we reduce the total uncertainty. Other improvements are superfluous.

One of the consequences is *that the accuracy (and, naturally, the price) of the measurement instruments do not need to be as high as possible.* The demand on those instruments that are responsible for the nondominant uncertainties can be reduced as long as the resultant uncertainties remain nondominant.

Conducting a measurement series reduces the *external* uncertainties of *this* series but, if the systematic uncertainty is dominant, this has no influence on the final result. Remember that *the total uncertainty is obtained by quadratic addition of the correlated and the uncorrelated uncertainty components* (Sect. 8.2.3), thus it does not make sense to make one of these two components a lot smaller than the other. Usually, the *precision*, given not only by the uncorrelated uncertainty but also by the numerical and the instrument resolution, and the *accuracy*, basically the systematic uncertainties, of the data should be well matched (see also Sect. 7.6).

Sometimes, it makes sense to improve the uncorrelated uncertainty even if the scale uncertainty remains large. This should, e.g., be done if regression analysis of the data values will be performed. In this analysis the scale uncertainty does not enter at all. Often such *relative* data are especially valuable in the context of extensive data evaluations.

Another possible way in which data evaluation and error analysis can influence the experiment is the following: If reliable data of experimental or theoretical nature are available for comparison, we can use our apparatus to "remeasure" these data. Then, e.g., by using a suitable graphical presentation we can identify (systematic) discrepancies. An analysis of these may lead to the detection and elimination of their origin. By applying the chi-squared test (Sect. 8.3.3) we can also check whether the uncorrelated uncertainties have been correctly identified.

10.1.1 Reasons For and Against Optimization

Optimization of a measurement, e.g., in a count rate experiment, should mean that the "best" result (i.e., the result with the smallest total uncertainty) is achieved in a given time interval, or that the required accuracy is reached in the shortest time possible. The following example, taken from real life, illustrates this point.

Example. *Overestimating Uncertainties*

In the course of data reduction the experimenter decided to multiply the (total) uncertainty (as determined by himself) with a factor of 3 "just to be perfectly safe". These numbers he then quoted in his publication.

Possible reasons for this:

- Too low self-confidence of the experimenter as far as the determination of uncertainty components is concerned because this subject is often neglected in education.
- Too much energy was used up for experimenting, so that the tedious work of correcting and of searching for the uncertainty contributions is passed over carelessly.

The consequences of such "modesty" can, in fact, be grave:

- If later on the same data are needed with a higher accuracy, the whole experiment must be repeated.
- If from the beginning a lower accuracy by a factor of 3 was meant to be sufficient, the measurement time could have been reduced by a factor of 9 (in such count rate experiments), i.e., almost 90% of the measurement time was wasted. For other types of measurements *all* measurement instruments used could be less accurate at least by a factor of 3 (and thus be cheaper by about the same factor), and the result would still have the same accuracy.

Optimizing a measurement usually means quite some effort on the part of the experimenter. It can be so demanding that it may cause actual errors to be committed by the experimenter. Therefore, it has to be decided from case to case whether an optimization of the experiment pays. Obviously, it does not pay if the time spent for the optimization is greater than the time gained by this optimization, e.g., by shorter measurement times.

10.1.2 Prevalent Design Criteria

If something that we know very little about is measured, such measurements are more or less a matter of luck. Sometimes they are the basis for those lucky cases where the theory must be revised or modified as a consequence.

In most cases experiments are conducted in the context of theories that allow an at least rough prediction of the result. Ideally, we could make the best possible choice of dependent variables based on these predictions. In reality things are slightly different; usually simple values, if possible integer numbers (in the units used), are chosen.

Example. ***Measuring Differential Cross Sections in the MeV Range***

Typically simple decimal energy values are chosen, if possible integer numbers, and angle values (starting from 0°) are chosen in steps of 5° or 10°.

The advantage of such a procedure lies in the easy direct intercomparison with earlier data (of the same reaction) if all experimenters proceed accordingly, i.e., the comparison is possible without the knowledge of energy and angle dependences. Additionally, remembering simple numbers is easier than consulting the measurement schedule when setting the required energy or angle values.

Another frequently found habit is the assignment of the same amount of time for measuring the fore- and the background, mainly because then the subtraction can be done directly with the same "live time" values (Sect. 4.1.3).

10.2 Optimizing Background Measurements

When measuring radiation spectra it is often necessary to subtract the influence of the background radiation (e.g., Sects. 3.2.6, 7.2.2, 7.3.2). Usually, this background spectrum is measured for the same length of (live) time as the foreground spectrum. However, if this time span is not of the order of a few minutes, at most, it is very advisable to optimize the partition of the total measurement time.

10.2.1 Optimized Simple Background Measurement

Let us assume that a total time t is available for measuring some gamma radiation in the presence of the natural gamma background. During the measurement of the foreground, we can approximately determine the ratio between the number of foreground events N_f and the number of background events N_b for the gamma line of interest from the spectrum recorded so far.

Assuming that the event rate ER is time invariant, we get

$$ER = \frac{N_f}{t_f} - \frac{N_b}{t_b} . \tag{10.1}$$

The optimum ratio of the measurement times t_f and t_b (yielding the most accurate result) is obtained from above equation by means of the *first derivative test* after applying the law of error propagation. From

$$(N_f/t_f^2) \cdot \mathrm{d}t_f - (N_b/t_b^2) \cdot \mathrm{d}t_b = 0 , \tag{10.2}$$

with the total time $t = t_f + t_b$ constant,

$$\mathrm{d}t = \mathrm{d}t_f + \mathrm{d}t_b = 0 , \quad \text{and consequently} \quad \mathrm{d}t_f = -\mathrm{d}t_b , \tag{10.3}$$

one gets

$$\frac{t_f}{t_b} = \sqrt{\frac{N_f}{N_b}} . \tag{10.4}$$

We can now determine both measurement times in relation to the total time:

$$\frac{t}{t_b} = 1 + \frac{t_f}{t_b} = 1 + \sqrt{\frac{N_f}{N_b}} . \tag{10.5}$$

Example. ***Splitting the Measurement Time***

While recording the foreground spectrum we find that the (flat) background spectrum in the region of interest is smaller than the foreground by a factor of about 20. The foreground count rate in the region of interest amounts to 100 events per minute. One hour is available as total measurement time.

- For measurement times of $t_{f1} = 30\,\mathrm{min}$ and of $t_{b1} = 30\,\mathrm{min}$, we get:

  $$N_1 = N_{f1} - N_{b1} = 3000 - 150 = 2850 \pm 56.1 , \quad \text{and} \quad \Delta N_1/N_1 = \pm 2.0\% .$$

- The same precision can be obtained for measurement times of $t_{f2} = 35\,\mathrm{min}$ and of $t_{b2} = 8\,\mathrm{min}$:

  $$N_2 = N_{f2} - N_{b2} \cdot 38/5 = 3500 - 175 = 3325 \pm 65.4 , \quad \text{and}$$
  $$\Delta N_2/N_2 = \pm 2.0\% .$$

For this example we have seen that the nonoptimized measurement time lasts longer by almost 40% than the optimized time under the condition of yielding *the same fractional precision!*

Problem

10.1. For the example of Sect. 10.2.1, to what percentage can the uncertainty of the result be improved if the total measurement time of the optimized measurement is also one hour?

10.2.2 Optimized Complex Background Measurement

It should come as no surprise that an optimization will be even more effective for more complex measurement problems. This can be seen in the following nontypical example from my own experimental work.

The cross section of the elastic scattering of fast neutrons from hydrogen is used as secondary standard for cross section measurements of fast neutrons, i.e., cross sections are often measured relative to the cross section of hydrogen (under similar or, if possible, the same conditions). Polyethylene is the usual choice as a solid hydrogen compound; it has two hydrogen nuclei for each carbon nucleus. Thus we not only get the scattering peak from hydrogen in our spectrum, but also one or more peaks from carbon. By measuring the scattering from graphite (carbon) to get the background spectrum we can subtract the carbon contribution to the polyethylene spectrum. Before this subtraction is done, both spectra must be corrected for the room background, which is the same in both cases. This room background has to be determined in another measurement without sample. Besides, the scale of the carbon spectrum has to be adjusted so that the size of the equivalent carbon peaks is the same.

To simplify the discussion let us assume that the neutron intensity stays constant in time so that the spectra can be normalized to equal time intervals. Let the count rates in the range of interest in the three spectra be N_P, N_G, and N_b, and the measurement times for polyethylene, graphite, and the background be t_P, t_G and t_b, respectively, then the reduced net count rate of neutrons scattered from hydrogen is given by

$$N = N_H/N_b = a_P \cdot (N_P/N_b) - a_G \cdot (N_G/N_b) - (a_P - a_G)\,, \tag{10.6}$$

where a_P is the correction factor for multiple scattering and attenuation in the polyethylene sample, and a_G is the combined correction factor for multiple scattering and attenuation in the graphite sample and the adjustment for the different amounts of carbon nuclei in the two samples.

The smallest fractional uncertainty in a given measurement time $t = t_P + t_G + t_b$ is achieved for

$$\frac{t_P}{t_b} = a_P \cdot \sqrt{\frac{N_P/N_b}{|a_P - a_G|}}\,, \tag{10.7}$$

and

$$\frac{t_G}{t_b} = a_G \cdot \sqrt{\frac{N_G/N_b}{|a_P - a_G|}}\,. \tag{10.8}$$

Thus, generalizing (10.5) all time ratios can be put in relation to the total measurement time as follows

$$t/t_b = 1 + t_P/t_b + t_G/t_b \,. \tag{10.9}$$

Compared to Sect. 10.2.1, where 80% instead of 50% of the measurement time is used for the optimized foreground measurement, the gain is even greater because without optimization only one third of the total measurement time is available for the foreground measurement. From experience we know that N_G as well as N_b can be pretty small, so that the same precision can be achieved in approximately one half of the total measurement time in the optimized case. So in most cases an approximate optimization will be adequate by choosing $t_P = 0.6t$, $t_G = 0.25t$, and $t_b = 0.15t$ instead of $t_P = t_G = t_b = t/3$; this would improve the precision by about 40%.

A more accurate procedure would consist in first choosing $t_P = 0.3t$ and then determining N_P/N_b from the measured polyethylene spectrum. The factors a_G and a_P can be determined beforehand with the help of Monte Carlo simulation and weighing. After having determined N_G/N_b in the following graphite measurement, we can now fix the measurement times. That is, the graphite measurement in progress can be stopped after the calculated time t_{Gopt} has passed, the background measurement can be conducted for the determined time length t_{bopt}, and then the polyethylene measurement can be continued for the remaining time of t_{Popt}.

10.3 Optimizing With Respect to Dead Time

The optimum count rate CR_{opt} gives the event rate ER with the smallest possible fractional uncertainty $\sigma_r = 1/\sqrt{N} = 1/\sqrt{ER \cdot t_m}$ within a certain time interval t_m. This optimum count rate depends on the dead time t_d of the detection apparatus. In some count rate experiments where the event rate ER can be set by choosing the appropriate experimental parameters (e.g., distance from the source, intensity of a particle beam), this circumstance can be employed. However, there are two types of dead time.

For *dead time of the first kind* the system is dead during a well-defined time interval t_d after detecting a signal, independent of whether other signals arrive during this dead time or not (like in Fig. 4.2). Obviously the maximum processing rate $CR_{\max 1}$ is

$$CR_{\max 1} = 1/t_d \,. \tag{10.10}$$

Any further increase of the event rate in the detector does not result in an increase of the (measured) count rate.

For *dead time of the second kind* each signal in the counting device, i.e., each event, causes dead time, not only signals that are processed. Thus we get:

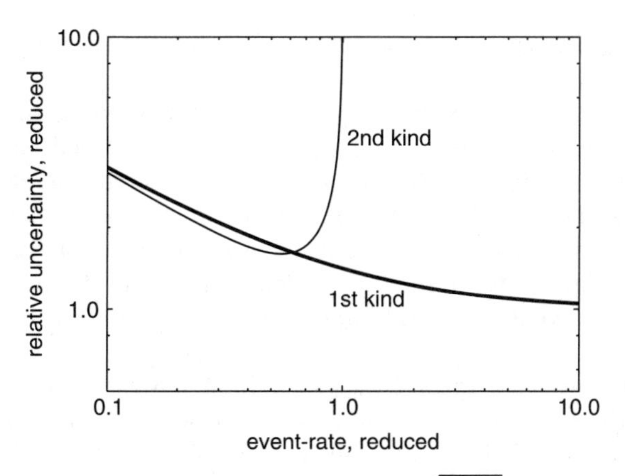

Fig. 10.1. Reduced relative uncertainty $\sigma_r(ER)/\sqrt{t_d/t_m}$ as a function of the reduced event rate $N_r = ER \cdot t_d$

$$ER_{\max 2} = 1/t_d\,. \tag{10.11}$$

At this input count rate (event rate) the detection system shuts completely off, resulting in no output signals at all.

In Fig. 10.1 the reduced relative uncertainty $\sigma_r(ER)/\sqrt{t_d/t_m}$ is presented as a function of the reduced number of events $N_r = ER \cdot t_d$.

For dead time of the first kind we get:

$$\sigma_r(ER) = \sqrt{\frac{t_d}{t_m}} \cdot \sqrt{\frac{1 + ER \cdot t_d}{ER \cdot t_d}} = \sqrt{\frac{1}{N}} \cdot \sqrt{1 + N_r}\,, \tag{10.12}$$

and in the reduced form, using the reduced number of events $N_r = ER \cdot t_d$,

$$\frac{\sigma_r(ER)}{\sqrt{t_d/t_m}} = \sqrt{\frac{1 + N_r}{N_r}}\,. \tag{10.13}$$

For dead time of the second kind we get:

$$\sigma_r(ER) = \sqrt{\frac{t_d}{t_m}} \cdot \sqrt{\frac{\mathrm{e}^{ER \cdot t_d}}{ER \cdot t_d} - 2} \Big/ (1 - ER \cdot t_d) = \sqrt{\frac{1}{N}} \cdot \sqrt{\mathrm{e}^{N_r} - 2 \cdot N_r} \Big/ (1 - N_r)\,, \tag{10.14}$$

and as above, in the reduced form,

$$\frac{\sigma_r(ER)}{\sqrt{t_d/t_m}} = \sqrt{\frac{\mathrm{e}^{N_r}}{N_r} - 2} \Big/ (1 - N_r)\,. \tag{10.15}$$

For zero dead time ($t_d = 0$ and consequently $N_r = 0$) the relative uncertainty of ER is identical in both cases:

$$\sigma_r(CR = ER; t_d = 0) = \frac{1}{\sqrt{N}}\,. \tag{10.16}$$

For $t_d \neq 0$ the difference between the two kinds of dead time increases strongly for $N_r > 0.7$.

For dead time of the first kind, the smallest relative uncertainty is obtained for $ER \rightarrow \infty$ as $\sigma_r(ER \rightarrow \infty) = \sqrt{t_d/t_m}$. For $N_r > 2$ only an insignificant decrease of σ_r with an increased count rate can be noted, therefore it does not make sense to increase the event rate ER any further than this $2/t_d$ value.

For dead time of the second kind, the smallest relative uncertainty σ_r occurs at a reduced event rate $N_r = 0.53$. However, the minimum is very flat, so that for event rates ER larger than $0.3/t_d$ no essential improvement is achieved.

The *optimum event rates* ER_{opt} of $2/t_d$ or $0.3/t_d$, respectively, are usually not aimed at because the count rate CR deviates so strongly from ER that the relative count losses are substantial (67% and 26%). Thus deviations from the mathematical model strongly affect the *calculation* of the dead time correction. With the help of the methods discussed in Sect. 4.1.2, even dead time corrections larger than 50% can be dealt with if done properly.

10.4 Optimizing in View of the Mathematical Presentation

Once the best mathematical presentation of the data is decided on, the best estimate, i.e., the parameters of the curve must be determined. In the following illustrative example, we will demonstrate that a least-squares fit using statistical weights, i.e., weights based on uncorrelated uncertainties, does not necessarily result in the best possible *best estimate*, i.e., in the best parameters representing the data set.

Example. ***Moessbauer Transmission Spectrum***

The simplest result of a Mössbauer transmission experiment (a velocity or energy spectrum) can be described by a Lorentzian curve. This curve is characterized by the position, the width, and the depth of its minimum, and by the zero resonance value (data values without resonance absorption). Usually a Mössbauer measurement is conducted as a multichannel measurement using velocity intervals of the same length; i.e., the velocity spectrum is described by equidistant points. Let us assume that the velocity steps are chosen in a way that the full-width at half maximum (FWHM) of the Lorentzian curve is smaller than the width of one channel; consequently the form of the dip is not resolved. It should be obvious that with the same number of points and the same precision, i.e., the same measurement time, the parameters of the above experiment will be much less accurate than if, for instance, the velocities had been chosen in a way that ten channels were equivalent to the FWHM. This is true even if a fit of the same (mathematical) quality (with the same sum of the squares of the deviations) is obtained in both cases.

The above example shows that the position of the points to be measured (the independent variables) should be chosen in a way that the parameters of the mathematical function can be determined with the smallest uncertainty possible. A disadvantage of such a preselection of the position of the data points is that we might overlook something interesting that is unexpected.

The field of optimizing measurements according to the shape of the mathematical curve, to my knowledge, has not been developed in a general way so far. Thus only some special cases will be presented here.

10.4.1 Optimizing Flat Dependences

Flat dependences mean that we are dealing with (weighted) means (Sect. 6.3.1). As far as the *uncorrelated (internal)* uncertainties are concerned, a measurement of *one* point twice as precisely is equivalent to measuring four points. This does not influence the correlated (systematic) uncertainty.

Thus it is better to measure as precisely as possible, rather than gathering as many data points as possible, except if we depend on external uncertainties.

10.4.2 Optimizing Linear Dependences

For linear dependences the main information usually lies in the slope. It is obvious that those points that lie far apart have the strongest influence on the slope if all points have the same uncertainty. In this context we speak of the strong *leverage* of distant points; when determining the parameter "slope" these distant points carry more effective weight. Naturally, this weight is distinct from the "statistical" weight usually used in regression analysis.

In Fig. 10.2 the effect of the leverage of a distant point, a so-called *isolated data point* is demonstrated. If the linear regression (minimum of the sum of the squares of the deviations) is calculated for the two data sets that differ only in the value of the first data point (i.e., one out of seven) we get two completely different lines. The right-hand line (b) has a positive slope even though six of the seven points (in the right region of the diagram) ask for a negative slope. One would expect that the linear correlation coefficient (Sect. 6.3.3) is quite different for the two regression lines. However, this is not so. Solution (a) results in $r_{xy} = -0.987$ and (b) in $r_{xy} = 0.954$. If these results were based on more data values (Sect. 7.4.2) they would even be significant. Although the absolute value for (b) is slightly smaller (as expected), the difference is not at all pronounced.

Note: Often the end points in distributions are not completely "understood", e.g., if they are near the upper or lower end of the operating range of the apparatus. In such cases correction factors might have been overlooked, resulting in wrong data values. In cases like in Fig. 10.2 this has grave consequences for the best estimate (the line) because of the leverage of the isolated end point.

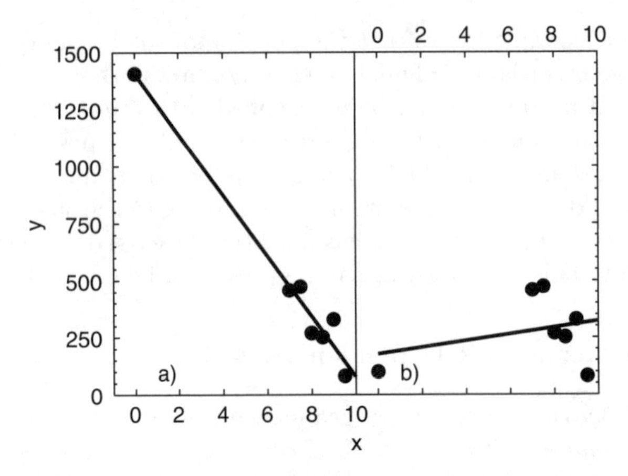

Fig. 10.2. Demonstrating the effect of the lever arm

10.4.3 Optimum Angles for Cross Section Measurements

After measuring differential cross section of a nuclear reaction the data values will be converted to the center-of-mass system. In addition, the parameterization is usually done via the sums of Legendre polynomials; therefore, it is not the angular dependence in the laboratory system that is relevant, but instead the dependence on the cosine of the center-of-mass angle.

A simple optimization of the angle would consist in equidistant steps of the cosine of the center-of-mass angle. Actually, the situation is much more complex, as not all data values are measured with the same accuracy because both the foreground and background count rates depend on the angle. Furthermore, one would need to know up to which degree the Legendre polynomials are needed for presenting the data properly before being able to determine the optimum angle values. It should be plausible that the impact of a data value on the best estimate is stronger, the fewer parameters contribute to this value.

Obviously, this procedure is far too complex to be accepted in practical work. Consequently, it is common to not optimize at all. Instead, the angles are usually chosen in equidistant degree steps modulo 10, 5 or 2.5 in the laboratory system (Sect. 10.1.2)!

10.5 Achieving the Smallest Overall Uncertainty

Let us summarize what we have learned so far about measures to keep experimental uncertainties as small as possible. The following two facts are of great help:

- Uncorrelated uncertainties are added quadratically, i.e., we must only minimize the dominant uncertainties (Sect. 3.4.1).

- Correlated (scale) uncertainties cancel entirely if they are present both in the numerator and denominator (Sect. 8.1.4).

Obviously, the best thing that can happen is that an uncertainty is canceled. Let us therefore discuss an experimental method in which as many uncertainties as possible are correlated and occur in ratios.

10.5.1 The Ratio Method

In the ratio method the measurements are not done relative to a primary standard, but relative to a secondary standard that allows for a direct comparison. Otherwise all provisions are made as for an absolute measurement. Therefore, this method is also called a quasiabsolute method.

First, the standard is measured using the experimental setup with which the actual experiment is conducted afterwards. This procedure can also be interpreted as calibration of the experimental setup by the secondary standard. After that, it is only important to keep the experimental conditions as constant as possible so that as little change as possible occurs between the measurement of the standard and the measurement of the sample in question. This way we ensure that as many uncertainties as possible remain practically identical, making them totally correlated.

Example. *Secondary Standard*

For a measurement of the differential cross section of the reaction ^{3}H(p, n)^{3}He at 13.600 MeV, many uncertainty components are listed in Table 10.1. For such a measurement, highly accurate reference data of the reaction ^{3}H(p,^{3}He)n are available obtained by measuring the charged ^{3}He particles. By a simple conversion into the center-of-mass system we get the desired reference data. Two possible sources of errors are common in such a transformation. First, we must remember that at the time of the nuclear reaction all particles are "stripped" of their electrons, i.e., the electron shells are empty because of the high velocity of the nucleus. Therefore we must use nuclear and not atomic masses for the kinematics calculations. Furthermore, the transformation from the laboratory system to the center-of-mass system has to be done relativistically because of the high velocities involved. In the experiment a proton beam of 13.600 MeV hits a tritium target producing monoenergetic neutrons by nuclear two-body reactions. The differential cross section is obtained by measuring the angular dependence of the neutron yield.

To probe the limitations of the ratio method, let us look at an especially simple measurement task where a maximum number of uncertainties are correlated. The aim is to increase the angular range of a (secondary) standard by a factor of about 2.

As shown in Table 10.2, the standard measurement (by Jarmie and Jett) only covers angles in the center-of-mass system between 77.20° and 155.50°.

Table 10.1. Uncertainty components when measuring differential cross sections of the reaction ^{3}H(p, n)^{3}He using the ratio method

Uncertainty type	Uncertainty component	Typical uncertainty
Unavoidable sources of uncertainties		
Systematic (or scale) uncertainties:		
Standard	Uncertainty of standard cross section	$\pm 0.7\%$
	Counting uncertainties	$\pm > 0.3\%$
	Center-of-mass conversion	$\pm < 0.1\%$
Uncorrelated uncertainties:		
New data	Counting statistics	$\pm > 0.3\%$
	Count loss correction	$\pm 0.2\%$
	Background subtraction	$\pm < 3\%$
	Relative neutron detection probability	$\pm < 3\%$
Systematic uncertainties of measurement parameters:		
Mean projectile energy		± 0.02 MeV
Angular zero-point position		$\pm 0.05°$
Uncorrelated uncertainties of measurement parameters:		
Individual angular setting		$\pm 0.05°$
Uncertainties that more or less cancel in a careful ratio measurement		
Normalization	Beam collection	
	Beam charge measurement	
	Beam heating correction	
Gas target	Areal density: pressure, length, and temperature	
	Isotopic abundance	
	Purity	
Counting	Background	
	Dead-time correction	
Detection	Counting efficiency	
	Geometric factors	
	Neutron absorption	
	Neutron scattering	
Detection angle	Zero degree reference	
	Position of detector	
Beam energy	Terminal voltage	
	Energy loss	

The use of neutrons instead of charged particles allows us to include angles in the forward direction. How can we extend the secondary standard to smaller angles, i.e., how can we get highly accurate data of differential cross sections $\sigma(E_p = 13.60\,\text{MeV}, \theta)$ for those laboratory angles θ that correspond to center-

Table 10.2. Absolute differential cross sections in the center-of-mass system for the reaction ^{3}H(p, n)^{3}He at 13.600 MeV

Angle (deg)	Jarmie and Jett (mb/sr)	$\Delta\sigma$ (mb/sr)	Best est. (mb/sr)	Drosg (mb/sr)	$\Delta\sigma$ (mb/sr)
0.00	–	–	21.315	$21.43 \pm 1.0\%$	0.214
13.50	–	–	18.982	$18.99 \pm 0.9\%$	0.171
26.76	–	–	14.125	$14.06 \pm 1.0\%$	0.141
40.11	–	–	10.356	$10.58 \pm 1.2\%$	0.127
52.90	–	–	9.406	$9.24 \pm 1.0\%$	0.092
65.58	–	–	10.293	$10.28 \pm 0.8\%$	0.082
77.20	$11.76 \pm 15.0\%$	1.76	11.350	$11.34 \pm 0.8\%$	0.091
87.67	$11.61 \pm 3.0\%$	0.348	11.541	$11.50 \pm 0.8\%$	0.092
98.04	$10.46 \pm 1.2\%$	0.126	10.520	$10.56 \pm 0.8\%$	0.084
108.37	$8.534 \pm 1.1\%$	0.094	8.552	$8.55 \pm 0.9\%$	0.077
118.65	$6.922 \pm 0.8\%$	0.055	6.881	$6.83 \pm 1.3\%$	0.089
128.93	$7.372 \pm 0.8\%$	0.059	7.365	$7.47 \pm 1.7\%$	0.127
139.17	$11.44 \pm 1.2\%$	0.137	11.545	–	–
149.37	$19.42 \pm 1.1\%$	0.214	19.473	–	–
155.50	$25.32 \pm 2.0\%$	0.506	24.943	–	–
Scale uncertainty	0.70%	–	0.80%	0.84%	–

of-mass angles $< 77.20°$? (*Note:* In this example the symbol σ stands for the cross section and not for the standard deviation.)

The number $N(E_p, \theta)$ of neutrons that are recorded by the detector at the laboratory angle θ with respect to the direction of the proton beam of energy E_p during a certain time interval t_m can be presented as follows:

$$N(E_p, \theta) = q \cdot N_t \cdot \sigma(E_p, \theta) \cdot \varepsilon(E_n)\,, \tag{10.17}$$

where:

- q is a measure of the number of protons hitting the target, based on the charge of the protons collected in the target during the time interval t_m.
- N_t is the *areal density* of triton nuclei in the gas target that is hit by the proton beam.
- $\varepsilon(E_n)$ is some kind of neutron detection probability for neutrons of the energy E_n.

Note: The interaction probability of protons with tritons in such an experiment is so low that we may use the linear approximation in the above equation.

As the charge measurement is used to determine the number of protons hitting the target (Sect. 2.1.5), we must ensure that really all protons are collected and included in the charge measurement. It is difficult to determine the effective areal density N_t reliably because the triton target is a gas target.

The effective target length must be known: the pressure in the target and its temperature, as well as the purity of the gas. These parameters are hard to determine also because the proton beam heats the target locally, causing a local decrease in density, depending on the beam intensity.

Many factors are included in the factor $\varepsilon(E_n)$:

- the solid angle under which the detector "sees" the target.
- the attenuation of the neutron intensity between the location of the neutron production in the gas target and the detector (in the gas, the target wall, in the air between the target and the detector, in the detector housing),
- the sensitivity of the detector material to neutrons; for the latter, the dependence on the neutron energy is of particular importance.

For an absolute measurement it would be necessary to know the uncertainties of all these components, and also the uncertainties of the mean proton energy and of the angular settings. In addition, the uncertainties of the calculated transformation from the laboratory system to the center-of-mass system and vice versa would have to be known.

When employing the *ratio method* two measurements are conducted under as identical conditions as possible

- the measurement of the standard:

$$N_S(E_p, \theta_S) = q_S \cdot N_{tS} \cdot \sigma_S(E_p, \theta_S) \cdot \varepsilon_S(E_{nS}), \tag{10.18}$$

  and
- the new measurement:

$$N_x(E_p, \theta_x) = q_x \cdot N_{tx} \cdot \sigma_x(E_p, \theta_x) \cdot \varepsilon_x(E_{nx}). \tag{10.19}$$

If the same target is used under the same conditions for the measurement of the standard and that of the sample the areal density cancels, and its uncertainty is irrelevant.

Provided that no changes occurred in the gas target during the time between the two measurements (so that the areal density is the same for both cases), we get $N_{tx} = N_{tS}$, and

$$q_x \cdot \varepsilon_x(E_{nx}) \cdot \frac{\sigma_x(E_p, \theta_x)}{N_x(E_p, \theta_x)} = q_S \cdot \varepsilon_S(E_{nS}) \cdot \frac{\sigma_S(E_p, \theta_S)}{N_S(E_p, \theta_S)}, \tag{10.20}$$

and further

$$\sigma_x(E_p, \theta_x) = \sigma_S(E_p, \theta_S) \cdot \frac{N_x(E_p, \theta_x) \cdot q_S \cdot \varepsilon_S(E_{nS})}{N_S(E_p, \theta_S) \cdot q_x \cdot \varepsilon_x(E_{nx})}. \tag{10.21}$$

In (10.21) the three factors N, q, and ε are found in the numerator as well as in the denominator, so that their correlated uncertainty components cancel.

Obviously, we must see to it that the experimental situation changes as little as possible between the measurements so that the fluctuating portion of the uncertainties of these three factors is as small as possible.

The following uncertainty components are inevitable:

- The uncertainty of the standard $\Delta\sigma_S$.
- The uncertainties ΔN_S and ΔN_x due to counting statistics, due to dead time corrections (partially correlated), and due to background determination (partially correlated).
- The uncertainty of the charge measurements Δq_S and Δq_x. Ideally, these two uncertainties are strongly correlated and cancel each other.
- The uncertainties of neutron detection $\Delta\varepsilon_S$ and $\Delta\varepsilon_x$. These have one correlated and one uncorrelated component. For a mechanically solidly constructed setup, the solid angle should change only minimally with the angle, thus this uncertainty component is correlated. The uncertainty in the corrections for the attenuation will also be strongly correlated. The energy dependence of neutron detection in the detector will have one correlated as well as one uncorrelated component. In cases where $E_{nx} \approx E_{nS}$ the uncorrelated component will be negligible, and this uncertainty will also cancel.

Obviously, the accuracy mainly depends on the accuracy of the standard and on the counting statistics of the standard measurement and the actual measurement. Apart from that, we must consider the effects of the uncertainties of the (mean) proton energy and of the (mean) angle. In Sect. 8.2.3 we discussed how the uncertainties of these two parameters can be included into the final result.

As for the angle, we must remember that the uncertainty of the angle consists of two components: the uncertainty of the zero direction, which is correlated (i.e., systematic), and the uncertainty of the measurement of the distance from the zero point, which will be chiefly uncorrelated.

In Table 10.2 the results of a measurement employing the ratio method are presented. Here things are slightly more complicated as six (secondary) standard data values are available instead of only one. That way, the uncorrelated uncertainty of the standard can be decreased (e.g., the counting statistics improved). The best estimate has been obtained by fitting Legendre polynomials. Averaging (using the weighted mean) of the nine ratios of best estimates to the standard values gives a mean ratio of 1.0000 ± 0.0039, where the uncertainty is the standard deviation of the weighted mean (Sect. 6.3.1). Although this uncertainty value is calculated using the *uncorrelated* uncertainties of the standard values, it is a *correlated (systematic)* uncertainty for all values of the data set. However, it is independent of the scale uncertainty of the standard of 0.70%, so that the scale uncertainty of the best estimate becomes $\pm\sqrt{0.70^2 + 0.39^2}\% = \pm 0.80\%$.

From the 12 ratios of the best estimates to the new measurement values an average ratio of 1.0000 ± 0.0027 is obtained as weighted mean. In this case the uncorrelated uncertainties of the measurement (mainly counting statistics) result in a correlated (systematic) uncertainty of $\pm 0.27\%$ obtained from the standard deviation of the weighted mean. This value is independent of the $\pm 0.80\%$ calculated above, so that we get a scale uncertainty of $\pm\sqrt{0.80^2 + 0.27^2}\% = \pm 0.84\%$ for the new measurement data. It is larger by only 20% than the $\pm 0.70\%$ of the standard.

Also, the size of the uncorrelated uncertainties of the data measured using the ratio method is comparable to that of the uncertainties of the standard. That the uncertainties of the new measurement are very close to those of the standard is the consequence of using the ratio method: Most uncertainty components have no effect because of the formation of correlated uncertainties.

Once again (like in Sects. 7.2.2, 7.2.3, and 8.2), we have seen that uncorrelated uncertainties can have systematic consequences; they can contribute to the (correlated) scale uncertainty.

This is only possible because correlated and uncorrelated uncertainties (or random and systematic errors) are of the same nature!

Problems

10.2.
(a) Apply the chi-squared test (Sect. 8.3.3) to the values of the standard and its best estimates in Table 10.2.
(b) Apply the chi-squared test to the measured values and its best estimates in Table 10.2.
Note: The best estimates of the 15 points were determined via a weighted least-squares fit of Legendre polynomials using 12 parameters in the wake of an energy-dependent evaluation (i.e., values of the cross sections of neighboring energies also contributed).
(c) How should the way the best estimates were obtained (see Note in (b)) affect the value of *chi*-squared?
(d) Name one basic reason why the credibility of the chi-squared test is less in (a) than in (b).

10.3. An underground mains line is short-circuited at some point away from the house. By measuring the resistance of the cable, the distance l_1 to the short circuit can be found. A 6½-digit multimeter is available with the following specifications: scale uncertainty $\pm 0.008\%$ (Sect. 3.1.1) and interpolation uncertainty $\pm 0.004\%$ (Sect. 3.1.2). The measurement, done in the 100-Ω range, gives $R_1 = 0.0972\,\Omega$. The specific resistance of the wires in the cable is known to be $0.00523\,\Omega/\text{m}$.

(a) Give the distance l_1 with uncertainties derived from this measurement.
(b) Give one possible reason for the short circuit to be closer than the measurement suggests.

(c) If the same type of cable is available and is short-circuited at the calculated length, its resistance R_2 can be measured, and a better estimate of the position l of the short circuit can be made: $l = l_1 \cdot R_2/R_1$. What would the uncertainty of l be?

10.4. A high-frequency data line is broken. To find the position of the interruption a signal from a fast-signal generator is fed into the line, and the arrival of the signal reflected at the point of interruption is measured by a high-speed oscilloscope. The travel time t_1 is measured to be 98.0 ns. The readout uncertainty of the two time marks is 0.4 ns each. The time calibration of the scope is within $\pm 3\%$. A spare cable of the same type and with a length of (10.000 ± 0.002) m is available. Measuring the reflection in this cable, one gets a travel time of 88.9 ns. At what distance is the line broken, and what is the uncertainty of this distance?

Solutions

Chapter 1

1.1
(a) 10^{-6}
(b) 10^{-10}

1.2 Theory

1.3 Uncertainties of numerical parameters and model intrinsic uncertainties

Chapter 2

2.1 Rotational velocity of a wheel

2.2 Revolutions per time

2.3 Distance per revolution

2.4
(a) Analog
(b) Digital
(c) Digital
(d) Analog
(e) Digital
(f) Digital
(g) Analog
(h) Digital
(i) Analog (because time is analog!)

2.5 2.252 – 2.3 – 2.25 – 2.3

2.6

(a) 34.
(b) 4.85
(c) 4.8
(d) 5.
(e) 0.06

2.7

(a) $8. \times 10^1$
(b) $7. \times 10^1$
(c) $4. \times 10^2$
(d) $4. \times 10^4$

2.8

(a) 32.35
(b) 32.7
(c) 32.4

Chapter 3

3.1 All best estimates have an uncertainty!

3.2 22.

3.3 46.2

3.4 2.1×10^7

3.5 Implicit	Explicit
(a) $10^{1.}$	$+21.62/-6.84$
(b) $1. \times 10^1$	$\pm 5.$
(c) 1.0×10^1	± 0.5
(d) 10.	± 0.5
(e) 10.0	± 0.05
(f) 10.000	± 0.0005
(g) $1/4 (=2^{-2.} \neq 0.25)$	$+0.10/-0.07$
(h) $(12.)_8$	10 ± 0.5

3.6 74 times

3.7 1.04%

3.8

(a) $(5.4 \pm 4.1)\,\mathrm{m}\Omega/\mathrm{m}$
(b) Nonlinearity of the ohmmeter
(c) Using a milliohmmeter the uncertainty could be reduced by about a factor of 10, or perform an optimized current–voltage measurement.

Chapter 4

4.1

(a) No
(b) Zero dead time

4.2 A good approach might be a drawing.

4.3 Statistical fluctuations

4.4

(a) 7.6
(b) 1.44

4.5

(a) 4
(b) 4
(c) 0
(d) 0
(e) 0%

4.6

(a) 0.101
(b) −0.055
(c) 1.094
(d) 0.346
(e) 343.%

4.7

(a) 6.36
(b) 0.78
(c) 11.8
(d) 3.7
(e) 58.%

4.8

(a) A data set with only one data point, a set of data where at least one value has a different weight
(b) A data set without a best estimate by way of a function

4.9

(a) No, only 3 (or at most 4) significant figures
(b) $9910. \pm 29.$

4.10 Graphic solution

4.11 From (4.25) one gets $a_0 = \sum y_i/n$ with $x_i = 0$.

4.12 $y = 95.$

4.13 For $x_m = 6.5$ one gets $y = -13.458 \cdot 6.5 + 9997.6 = 9910.1 = y_m$.

Chapter 5

5.1 Divide all data values by $n = 100$

5.2 2.77

5.3

(a) 9913.3
(b) 9910.1
(c) Loss of resolution because of the grouping into classes

5.4

(a) 171.4 cm
(b) 171.43 cm (assuming equidistant height distribution in the bin)
(c) 170.5 cm
(d) 30.0 cm
(e) 2.6 cm
(f) 1.5%

5.5

(a) $0.77 \neq 3 \cdot (-0.23)$
(b) $-29.9 \approx 3 \cdot (-6.9)$
(c) $0.9 \neq 3 \cdot (-0.03)$
(d) Too few data points, wide bins

5.6

(a) Mode
(b) Median
(c) Mean

5.7 $1. - 0.9^5 = 0.41$

5.8 37.%

5.9 $1! = 1 \times 0! = 1$

5.10 5.0%

5.11 A line parallel to the y-axis with a height of 5168 at the position $x = 16$

5.12

(a) 0.135×10^7
(b) $(1. - 0.14 - 0.27 - 0.27 - 0.18) \cdot 10^7 = 1.4 \times 10^6$

5.13

(a) 1.00σ
(b) 2.00σ
(c) 3.00σ

5.14 $0.5 \cdot (0.8740 - 0.5878) = 0.1431$

5.15

(a) 683
(b) 253
(c) 23
(d) 136

5.16 15.9%

5.17 1/2

5.18 3/4

5.19 1/2

5.20 11/36

5.21 1/9

5.22

(a) 1/169
(b) 1/221

5.23 16/52

5.24 15.2%

5.25

(a) 0.25
(b) 0.75

5.26 $p_S = 2/11$

5.27 $p_1 = 3/13$, $p_2 = 3/13$, $p_3 = 4/13$, $p_4 = 2/13$, $p_5 = 1/13$

5.28 7.6%

5.29 No, answer is within statistical uncertainties.

5.30 For example, the ratio of female to male shoppers

5.31 The ratio of the average length of time one has to wait for either train.

5.32

(a) 0.0063%
(b) 39.3%

5.33

(a) $(175.0000 \pm 0.0043)\,\mathrm{cm}$
(b) It is the best estimate of what a Chinese *thinks* that the height of an American is.
(c) The data values are not normally distributed.

Chapter 6

6.1

(a) $y_m = 9910.1$
(b) $\sigma = \pm 28.7$

6.2

(a) No
(b) The data are not normally distributed.

6.3

(a) 100.1 vs. 99.5
(b) 28.7 vs. 28.9

6.4 (b) and (d)

6.5 $\Delta y_{mw} = 1/\sqrt{(1/\Delta y_1)^2 + (1/\Delta y_2)^2}$

6.6 $\Delta y_m = \Delta y/\sqrt{2}$, with $\Delta y = \Delta y_1 = \Delta y_2$

6.7 0.58

6.8 6.0

6.9 Nondominant components of weighted means may not be disregarded.

6.10 $y_m = (9910.1 \pm 28.7)$ events per minute

6.11 $a_1 = -13.458$, $a_0 = 9990.9$

Chapter 7

7.1 Age

7.2

(a) $\Delta V_m/V_m = \pm 0.022\%$
(b) $\Delta V_m/V_m \approx \pm\sqrt{3} \cdot 0.02\%$

7.3

(a) Ratio $R_s = 1.951 \pm 0.086$
(b) Because the uncertainties of the foreground counts are dominant

7.4 See the *transmission line* example

7.5

(a) $N = 11,111$ events, $ER = 1111$ events/s
(b) $\Delta ER/ER \approx \Delta N/N = \pm 0.95\%$. The uncertainty of the time measurement can be disregarded.

7.6

(a) $N = N_t - 0.5 \cdot (A_1 + A_2) \cdot [X_2 - (X_1 - 1)] = 14{,}512{,}310$
(b) $\Delta N = \sqrt{(N_t + 0.25 \cdot 104^2 \cdot [A_1 + A_2])} = \pm\,17{,}863$.
(c) $\Delta N/N = \pm\,0.12\%$ vs. $\Delta N_t/N_t = \pm\,0.02\%$, i.e., the uncertainty of the background is dominant.

7.7

(a) $N = N_t - N_{bg} = 18,582,987$
(b) $\Delta N = \sqrt{N_t + (0.2 \cdot N_{bg})^2} = \pm 335{,}199$
(c) $\Delta N/N = \pm 1.8\%$ vs. $\Delta N_t/N_t = \pm 0.02\%$, i.e., the uncertainty of the background is dominant.
(d) 7.6, uncertainty of the background is $\pm 0.3\%$; however, the more sophisticated background of 7.7. is 71% smaller!
(e) 7.7, despite the fact that the uncertainty of the background is assumed to be $\pm\,20\%$, the best estimate is expected to be much closer to the true value than in Problem 7.6.

7.8 See Sects. 7.4.1 and 7.4.2.

7.9 Yes

7.10

(a) Yes, $r_{xy} = 0.934 \approx 1$.
(b) $a_0 = 73.72$, $a_1 = 1.327$

7.11 1.34%

7.12 $0.89\% < 1\%$, highly significant

Chapter 8

8.1

(a) $\Delta F = x^{m-1} \cdot y^{n-1} \cdot \sqrt{(m \cdot y)^2 \cdot (\Delta x)^2 + (n \cdot x)^2 \cdot (\Delta y)^2}$
(b) $\Delta F/F = \sqrt{m^2 \cdot (\Delta x/x)^2 + n^2 \cdot (\Delta y/y)^2}$

8.2

(a) Averaging over slightly different diameters
(b) $65.5\,\mathrm{mm}^3$
(c) $2.0\,\mathrm{mm}^3$

8.3 $\Delta\varphi = \tan(\varphi) \cdot \sqrt{(\Delta x/x)^2 + (\Delta X/X)^2}$

8.4

(a) Graphic solution
(b) The line intersects five uncertainty bars instead of the expected four
(c) $chi^2 = 0.69$
(d) The correlated component is 65.8

(e) $chi^2 = 1.00$
(f) Trustworthiness of the chi-squared test requires at least ten degrees of freedom.

Chapter 9

9.1 $\pm 0.012\,\mathrm{V}$

9.2 Seven times shorter; compare ± 0.150 with ± 0.055

9.3 See Sect. 9.3

9.4
(a) 10^{-6}
(b) Temperature variations

9.5 Example 3 in Sect. 7.3.2

Chapter 10

10.1 $\pm 1.7\%$

10.2
(a) 0.27
(b) 0.73
(c) The data are overfitted.
(d) Fewer independent data, less than ten degrees of freedom

10.3
(a) (9.29 ± 0.38) m
(b) If the short circuit has a resistance > 0.
(c) Negligible. The scale uncertainties cancel and the interpolation uncertainties nearly cancel because both measured values R_1 and R_2 will be very close to each other.

10.4 (11.02 ± 0.09) m because scale uncertainties cancel.

Index